Sept 2012

# BACKGARDEN CHICKENS
# AND OTHER POULTRY

# ABOUT THE AUTHORS

John Harrison, bestselling author of *Vegetable Growing Month by Month*, has kept chickens for many years, originally in a small back garden, now on his smallholding in Wales. His daughter, Cara Harrison, keeps her family of chickens, ducks and quails in the back garden of her terraced house and blogs about them on the UK's number 1 allotment website: **www.allotment.org.uk**

*Other books written by John Harrison,*
*available in the Right Way series:*

Vegetable Growing Month by Month
The Essential Allotment Guide
Vegetable, Fruit and Herb Growing in Small Spaces
The Complete Vegetable Grower
Low Cost Living
How to Store Your Home Grown Produce*
Easy Jams, Chutneys and Preserves*

*with wife Val Harrison

# BACKGARDEN CHICKENS AND OTHER POULTRY

John and Cara Harrison

**RIGHT WAY**

Constable & Robinson Ltd
55–56 Russell Square
London WCIB 4HP
www.constablerobinson.com

First published by Right Way,
an imprint of Constable & Robinson, 2011

A copy of the British Library Cataloguing in
Publication Data is available from the British Library

ISBN: 978-0-7160-2268-8

Printed and bound in the EU

1 3 5 7 9 10 8 6 4 2

# CONTENTS

Our thanks to Stephen Liddle
for his illustrations

# ILLUSTRATIONS

# COLOUR PLATES

# INTRODUCTION

*To close one's eyes and dream of a home in the country with its lawns, its gardens, its flowers, its songs of birds and drone of bees, proves the sentimental in man, but he is not practical who cannot call into fancy's realm the cackle of the hen.*
From: *Making a Poultry House* by M Roberts Conover, 1912

This book is for those considering or already keeping some poultry in their back garden. Although most people start with chickens, we've also touched on the other poultry that you could have: ducks, turkeys, geese and quail.

Ducks, in particular, are enjoying a resurgence as a backgarden bird. Despite what you may think, you don't need a pond and we explain how to keep happy, healthy ducks without one. They mix well with chickens in the same garden so there's no reason not to enjoy both.

A book is in some ways a reflection of its authors and this one is no exception. Cara Harrison keeps her hens, ducks and quail in a small urban back garden while her father, John, has a smallholding in Wales after many years of suburban living and keeping an allotment.

Both of us aim to be as self-sufficient as reasonable in the modern world and our approach reflects that. But that doesn't mean we're totally practical, as our collection of retired aged hens and pet ducks who escaped their destiny in the oven will show.

This is not a book for those who want to make a living from

their poultry but we show how you can, if not cover, at least make a contribution to the cost of your hobby from selling your eggs.

Many people simply keep their poultry as pets who give them eggs as a bonus, and that's perfectly fine. It's often a surprise to new keepers just how responsive and entertaining chickens and ducks can be. Not to mention how therapeutic it is to discuss your problems with your flock. They might not answer, but they're great listeners!

On the other side of the coin, more people are now considering raising their own table birds. When you look at the horrific reality of large-scale commercial production of poultry meat, there is a lot to be said for raising your own. Not just on humanitarian grounds either, we explain why poultry meat home-raised on a small scale is actually better for you. Raising table birds isn't for everyone and we explain the pros and cons so you can decide for yourself.

We've tried to cover everything you might need to know when keeping a small flock at home but also tried to avoid swamping you with too much information. For example, we cover the most popular breeds for those starting out but not every breed. In fact, you could fill an entire book with just the different breeds of chicken you can find. And end up so confused you'd never get started!

We'd like to thank Lesley Williams and Ann Clarke for their help and criticism of the first draft of this book. We'd also like to thank the members of our website forum for their help: the beginners who asked the questions we would never have thought of and the experts who share their knowledge so freely.

# 1

# PREPARING FOR YOUR POULTRY

Before setting out to collect your first birds you should make sure that you have everything ready in advance. Trust us: it will make life easier for everyone if you are well prepared.

One of the first things to do is to find a local vet who is willing and capable of treating poultry. Not all small-animal vets will treat them and some will treat but have not done the relevant courses to know enough about poultry-specific illnesses.

Asking around other poultry keepers or checking online lists of poultry-friendly vets is a good idea before you get your first birds. Although a lot of poultry illnesses can be home-treated, some will likely need veterinary treatment and, if you yourself are unwilling to cull, you will at some point need the vet for that sad final injection.

Chickens, ducks, geese, turkeys and quail may not be family pets for some people but they do still deserve a high level of care and should never be left to suffer. Usually when you need a vet it is an emergency and knowing which vet you can call will be a boon.

Going on a backgarden poultry-keeping course is a good idea for those who have never kept poultry before. We hope this book will give you a good grounding and tell you all you need to know to get started, but no book can compete with actually handling a bird.

In addition, if you intend to breed meat birds it is a good idea to attend a course where safe culling methods are taught. If you wish to treat your own birds and come the final day do the deed yourself instead of using a vet, you should also make sure you know how to humanely cull an injured or sick bird. We would, in fact, recommend that all poultry keepers learn how to humanely cull. We have had occasion to do this late at night and on a weekend when the vets were not open and when leaving the bird to suffer would have been, in our opinion, cruel and inhumane.

Setting aside a poultry store cupboard whether in the house or shed is a good idea. You'll gradually build up a stock of those things you need occasionally and keeping them in one place means you'll easily find them.

The checklist below covers what you need to have ready before you bring your birds home – or what you are going to need pretty sharpish if you've arrived back home from a show with three hens in the back of the car!

**Basic New Poultry Keeper Checklist**
**Housing** – Make sure you have a poultry house that is the right size and style for your new birds.

**Run** – If you have a predator problem or wish to restrict the area your new birds are kept in, a run should be in place before they come home to you.

**Bedding** – You need to have decided which bedding and nest box filler you are going to use and have got this prior to collecting the birds.

**Feeder and Drinker** – You need to have both of these ready for your poultry and will have to make sure they are large enough for your new flock. Having a spare feeder and drinker is also a good idea.

**Feed** – Make sure you have the right feed for your new birds. Often the breeders will tell you what they have been feeding the birds on before you collect them.

**Disinfectant/Cleaning Equipment** – You will need to clean the house at least once a week and it is also a good idea to do this before the birds come home with you. Dry disinfectants are better for winter months when wooden poultry houses will take longer to dry out.

**Egg Storer** – You will need somewhere to keep the lovely eggs your new girls are going to give you, so make sure you have somewhere to store them, either a cool, dark place or in the refrigerator. Once you have refrigerated eggs, you will need to keep them in there. This is due to eggs being easily susceptible to temperature changes as well as bacterial development increasing at room temperature thereby reducing egg quality and life span if you swap them back.

In time, new additions will be made to your poultry store cupboard, including items such as wormers, red mite powder, lice treatments, Apple Cider Vinegar, purple antiseptic spray, beak bits, etc. You may wish to get these in advance or wait and get them as your flock needs them.

# 2

# OTHER PETS AND CHILDREN

Don't forget that your poultry are going to be a big part of your life and have an impact on the whole family. You'll need to consider how to manage this to keep everyone happy, healthy and safe.

**Children**
Most children love helping to look after family pets and this applies just the same to keeping backgarden poultry. They'll delight in finding those eggs that have magically appeared and bringing them up the path for breakfast.

Chickens are the most common choice of poultry to keep with a young family. Some breeds are calmer and less flighty than others and hybrids, in particular, tend to be calmer with young children. It's a source of wonder to us that ex-battery hens especially can become so attached to the humans in their life after being so mistreated.

Regular handling of the birds and teaching children how to hold a chicken safely and securely will help them to bond with their new pets. Getting the children to give the hens their afternoon treats also helps the bonding process and to gain the hens' trust. Hens will rarely attack the hand that feeds them, although they can become over enthusiastic if they think there are treats on the way. Sometimes they will accidentally peck which can, with younger children in particular, hurt a little, although some say it just tickles.

If you intend to breed and want to keep a cockerel running with your flock, you will need to make sure you choose a good-natured bird as they can, especially during the breeding season of early spring and summer, be very protective of the hens. Keeping an eye on younger children around cockerels is a good idea as if they attack they can cause injury with both beak and spurs. In fact, we'd suggest that unless you are really certain of a cockerel's temperament, it is better not to keep one when young children may get in with him.

Ducks are another good choice, more so with older children. They tend to be skittier than chickens and are less likely to be easily held, which is of course a major enjoyment for children with their pets. Ducks will rarely cause accidental injury as they don't tend to peck in the same way as chickens and they enjoy being hand fed which is enjoyable for children and adults alike.

As with cockerels, come the breeding season you need to be careful if running a drake alongside your ducks. They can be very protective of the girls in their flock. A bite from a drake can hurt, especially if a heavier breed. Young children should always be supervised if you have drakes to avoid injury.

Remember that chickens and ducks do not need a male counterpart in order to lay. You only need to get one if you want to breed them. It's best to start with just females if you have children. Allow the children to become bonded and feel safe around their new pets before you expand the flock.

Geese, in particular, will need a pond. Obviously, with younger children, ponds may be a concern in case they accidentally fall into them. It is safest either to make sure that the pond is beyond a gated area so the children can only get to it supervised or not to keep geese until the children are older.

Geese have a reputation for being aggressive which is not totally deserved but nevertheless we'd not recommend mixing younger children with geese.

Quail are enjoyable birds to watch but, as flight birds, need to be kept within a run which reduces the fun interaction children have with them. Older children can be taught how to handle the quail properly and younger children will be entertained by watching them flutter about their run, but children will experience greater enjoyment from keeping chickens, ducks or geese.

Whatever the type of poultry, children will need to know not to chase or frighten the birds as this will make them nervy not only of children who come near, but also of adults; the birds may also attack if scared. When a hen, duck or goose is broody, she can also become territorial, so children should be kept away from mothers and their young for the first few weeks. If a child wants to hold the hatchlings, this should always be supervised to make sure not only that the hatchlings are safe but that the child is safe from the mum who will want to protect her brood.

One final note on keeping poultry with children which, in fact, applies to adults as well: be aware that, although not a common problem, there are some diseases that poultry may pass on to humans. After handling your birds or cleaning them out, you should always give your hands a thorough wash with antibacterial handwash. This is particularly important for children and pregnant women. Using antibacterial soap and water is far better at removing lurking nasties than using an antibacterial liquid gel.

## Dogs

Dogs are a common family pet and one that needs to be handled carefully when getting poultry for the back garden. As a dog owner you may feel you know that your dog is safe with other animals, but attacks by dogs on new poultry are very common and a careful introduction should be carried out.

You should always keep your dog on a lead when he's with the new birds and test his reaction to them to begin with. Allowing the dog to see the birds and the birds to see the dog will help them both get accustomed to each other. If the dog appears to be safe with the birds, you can try supervised sessions off the lead to see how he interacts with them.

Remember that if the birds run at speed or fly, the dog's natural prey reaction may kick in and he may chase the birds. Although the dog may not intend to hurt them, accidents can happen and an overly excited, playful dog can accidentally cause as much damage as a dog that intends to attack them.

It's difficult to know how an individual dog will react to these new inhabitants in the garden. Some dogs will be jealous of the attention the new arrivals get and others may see them as a threat to the family, especially to the children. Others will act like a sheep dog, herding and trying to round up the hens for their owner.

If you feel that your family dog may, accidentally or due to its natural prey reaction, harm the birds, you will need to make sure that the dog cannot get to the birds, either through placing them in a secure run or keeping the dog away from the area where they range freely.

Having a dog should not put you off keeping poultry. Many dogs become intensely protective of "their" flock and will help to herd them to bed and watch out for other predators. Over time and with training, most dogs will be able to be left with the poultry, but by taking time and introducing them slowly, you make sure that they are happy and safe with your new birds.

## Cats

Cats have a natural instinct to hunt and most cat owners will have had a present dropped on the floor of a mouse or a bird that they have just caught. This makes people worried that keeping poultry with cats may not be a good idea and that neighbourhood cats may also be a threat to their birds. In fact, pet cats are far less of a threat to poultry than pet dogs. They are often quite terrified of the giant sparrows taking over their garden!

If you have cats or know that there are a few in the neighbourhood, your main consideration should be what size of bird to keep. Bantam chickens are at the most risk from cats as they are small enough to be seen as prey. Despite having many prolific hunters in our house, we have never had a problem with them or with any other neighbourhood cats attacking our bantams. However, we know of others who have, so if cats are around, a run for smaller birds is the safe option.

Birds from a hybrid size up are fairly safe from cats as predators. When we first got three hybrid chickens, the cats wouldn't dare come within five feet of them. If the chickens approached, they ran, tails fluffed back, to the safety of the house. As with dogs, introductions should be watched carefully for the first week or two just to make sure they are fine sharing the garden together with this system. We have on more than one occasion found one of the cats penned in a corner by the chickens desperate to escape the monsters. When it comes to the heavy breed ducks, even after a number of years sharing the garden, they seem to feel a five foot circle should be made to avoid them!

Small flight birds such as quail, however, are not safe from cats. You will need to make sure any house and run are cat-proof as most cats will show interest in getting into the run and they will kill them. We find that even now the younger cats will watch the quail intently and we know that if we don't lock them up securely we will wake up in the morning to no quail. The quail do not seem overly concerned by the cats watching them. So as long as they are secure, which they need to be anyway, keeping quail and cats should not be a problem.

As with quail, young hatchlings are not safe from cats. Although the mother hen or duck will protect their young, this may not be enough to deter a persistent moggy who has spied the hatchlings. Once larger breed chickens and other hatchlings have reached 8 weeks, they should be big enough to be allowed to range within an area. Younger hatchlings should be either secure in a run with mum or supervised when out in the garden.

Don't forget: you cannot train a cat to leave young birds alone. The best you can hope for is to train them to leave them alone whilst you are watching. When you're not watching, they're fair game to the cat.

### Rabbits

Many people will keep pet rabbits in their garden and there are many poultry keepers who use the same run for their chickens and rabbits with no problems. If you do intend to allow them to range together, you need to make sure they have separate housing and that, if they are within the same run, the rabbits are not persistent diggers. If the rabbits are always digging and are kept within a run designed to keep them safe from predators, the rabbits' digging may weaken the run and create entry points for predators such as foxes or stoats. We don't run our rabbits with our chickens and ducks, but this is a case of personal choice.

## Fish in Ponds

If you keep ducks and have ornamental ponds with fish in them, you will want to make sure that the ducks do not get to these. Not only will they cause damage to the plant life, they are likely to kill any fish in there.

# 3

# HOUSING, RUNS AND EQUIPMENT

*Regard it as just as desirable to build a chicken house as to build a cathedral.*

Frank Lloyd Wright

## HOUSING

Chickens, ducks, geese, turkeys, quail and all other poultry share a need for safe, warm housing that is free from draughts and predator-proof. There are so many options that it can be quite daunting to decide what is best for you and your birds.

Should you choose the traditional wooden chicken coop, a combination house and run, a plastic house or ark, or should you go for a self-build or converted shed? All are good options, but what matters is making the right choice for you – for the amount of land you have available, for the number of birds you want, your building abilities and for a price you can afford, of course.

When deciding on housing, it is best to sit down, pen and paper in hand, and compile a checklist of needs and priorities so you know just what you are looking for.

**Budget** – How much do you have to spend? This budget is just for the housing. The feeders, drinkers, hygiene products, health products, bedding and feed will all be on top of this.

**Space** – How much space do you have available? Remember that even if you can fit a coop in that sleeps eight chickens,

21

do you have enough run space for all these hens to roam in?

**Type** – What type of housing do you want? If you want a coop and run all-in-one, make sure you have space for the run and possible run extensions if you decide to expand your flock.

If you are handy and intend to build your own, make sure you have a plan. Ideally view other people's homemade coops so that you create a palace for your chickens and not a shack.

Make sure you know what features are important to you. Pull-out droppings trays, easy-access nest boxes and removable perches all make your life as the poultry keeper much easier, even though the poultry themselves won't care if these additions are in place or not! Don't forget that chickens produce the bulk of their droppings at night in their sleep, so ease of cleaning is very important.

## Space
In Britain and the EU there are legal stocking densities for poultry depending on the rearing method. In response to public disquiet – well, horror and anger may be a better description – the legal requirements for battery caged hens were changed to an enriched system with more space per hen.

Legally in the EU a laying hen in a post-2003 installed enriched cage is given $750 \text{ cm}^2$, a nest, a 15 cm perch and a scratching area. Many cages are still the pre-2003 enriched cage and these have just $550 \text{ cm}^2$ per bird – little more than a sheet of A4 paper per bird. Currently (2011) it is planned to make the enriched cages the only method of caging hens in the EU as of 1 January 2012.

Even the new enriched system hardly provides for the happiness of the hens. The terrible condition of newly rescued caged hens just proves to us how bad the legislation and conditions are for millions of hens.

The barn system may sound better, the hens being kept in a loose flock usually with access to different levels, perches, etc., but the stocking density allowed is actually higher than enriched cages at 9 hens per square metre! The barn system, with perches and feeders at different levels, gives a litter area of 250 cm$^2$ with 15 cm of perch per bird and one nest for every 7 birds.

In the free-range system, the birds are housed in the same way as the barn system, but in addition they are given continuous daytime access to open runs with 2,500 birds per hectare. This works out to 4 m$^2$ per hen in theory.

In practice, with the large flock size of commercial free-range poultry keeping, the hens don't actually tend to wander too far from the barn, so the real stocking density is far higher on the pasture next to the house. In fact, many of the hens in large-barn free-range systems never actually make it outside from the vast buildings called a barn.

All of the above goes to show that as guidance for the home poultry keeper, the EU stocking density regulations are basically irrelevant. Of course, within reason, the more space you give your birds, the happier they will be and the better the rewards for you both in terms of egg production and the enjoyment of watching them in the garden enjoying themselves. Unlike the commercial producer, your birds' happiness is a prime factor for you to consider.

The following table shows the minimum and recommended space in terms of runs and housing for the most popular egg-laying or pet poultry:

| Type of Poultry | Minimum Housing per bird | Ideal Housing per bird | Minimum Run per bird | Ideal Run per bird |
|---|---|---|---|---|
| Chickens | 1 ft$^2$/0.1 m$^2$ | 2 ft$^2$/0.2 m$^2$ | 10 ft$^2$/1 m$^2$ | 20 ft$^2$/2 m$^2$ |
| Ducks | 2 ft$^2$/0.2 m$^2$ | 3 ft$^2$/0.3 m$^2$ | 16 ft$^2$/1.5 m$^2$ | 32 ft$^2$/3 m$^2$ |
| Geese* | 3 ft$^2$/0.3 m$^2$ | 6 ft$^2$/0.6 m$^2$ | 32 ft$^2$/3 m$^2$ | 64 ft$^2$/6 m$^2$ |
| Turkeys | 3 ft$^2$/0.3 m$^2$ | 12 ft$^2$/1.1 m$^2$ | 40 ft$^2$/3.7 m$^2$ | 64 ft$^2$/6 m$^2$ |
| Quail | 0.5 ft$^2$/0.05 m$^2$ | 1 ft$^2$/0.1 m$^2$ | 1 ft$^2$/0.1 m$^2$ | 2 ft$^2$/0.2 m$^2$ |

*Geese require grass which is not included in the run figure.

The minimum recommended space is the very least you should be giving your birds to keep them happy, healthy and comfortable. Giving them the ideal space in both house and run will help minimize boredom and stocking problems such as feather pecking and cannibalism.

**Where to Put the Coop**
Often coops and runs are moved around the garden to allow access to fresh vegetation and prevent an area becoming over-used and damaged. However, if you are keeping them in the same place there are a few points to think about.

Avoid hot spots and direct sunshine if possible. Some shade will help keep the coop cool in hot sun; heat can be a killer.

In windy areas, position the coop in a sheltered place or at least turn it so the prevailing wind isn't whistling through the door.

Avoid wet areas. Damp coops are a breeding ground for disease and wood will rot quickly.

**Traditional Wooden Chicken Coops**
Until recently all chicken coops were constructed from wood but now you can obtain plastic houses (which we cover later). These have some benefits over wood. However, wooden houses can look very attractive and many people dislike the garish colours and artificial appearance of some of the plastic houses. Of course, there is nothing to stop you painting the outside of your coop if you wish. We've seen some that are works of art.

**Buying a Coop**
There's a huge range of ready-made coops to buy nowadays, especially on the internet, which makes the job of choosing very difficult. As with most things in life, the general rule is you get what you pay for.

Many coops are made in China and imported. They look very attractive and the pricing is extremely competitive. The manufacturer's price can be as low as £50 – which after freight, taxes and retailer's mark-up means they can retail for £120.

They look like incredible value but sadly they aren't. The designs can be very good overall, but everything is trimmed to the bare minimum. The wood is thin, the screws cheap and metal parts like hinges just adequate. They just don't last: one knock and they're broken.

As a general rule, it is better to buy wooden coops made domestically and, if at all possible, see before you buy. Even an expensive price isn't a guarantee of quality or design. Check the thickness of timber. It should be solid enough to take the inevitable knocks and bumps. It should be seasoned so unlikely to warp and twist, and should be treated with bird-friendly preservative to prevent rot.

Often you will see chicken coops marketed for "8–10 large birds" or "12–14 bantam-size birds" but you need to be careful as often these descriptions can be misleading. When buying a coop make sure you find out the floor area of the sleeping unit, the length of perches, and size and number of nest boxes. Often coops marketed for 8 birds are, in fact, only 6 ft$^2$ (0.6 m$^2$) in the sleeping area and as such should not sleep more than 6 birds, ideally only 4 large fowl.

When purchasing a coop, you need to make sure that it is easy to clean. Most people assume this to be an obvious fact, but you will be amazed by how many coops are difficult to get into to clean out properly, with badly positioned or too small doors or without lifting roofs to allow easy access.

Keeping the coop clean is essential to prevent red mite and other parasites affecting your birds so a coop that can be cleaned in all corners is a necessity. Beware felt roofs. These provide a great hiding place for mites. Roofs made of solid wood, external plywood or material such as Onduline corrugated sheet are far superior.

The coop will need to have good ventilation above the head height of the birds. Ventilation is important as fresh air is needed to circulate at night, but draughts can kill chickens so the ventilation must be at the right height and ideally vents will be adjustable to allow greater airflow in the hotter summer months.

Many owners are concerned about their birds being cold in winter but they're actually more at risk from summer heat. Don't forget that poultry come with their own feather duvet.

Unless the winter temperature is below −5°C (23°F), the birds will be fine in the house with their own body heat. Below that temperature you need to consider some additional heat.

Carrying handles are a good addition if you intend to move the coop around a garden or field. It is also useful to move the coop to clean under and around it so, even if you want to keep the coop in a fixed place, handles can prove to be beneficial.

The height of the coop and whether or not it is raised off ground level should also be considered. If the coop is off the ground, make sure it is by more than 2 inches (5 cm) so that you can see under the coop. Often rats and mice will use gaps to access the coop (by gnawing their way in) so a good view of the base of the coop is beneficial to prevent this. Similarly, if your coop is going to be sited on soil, remember that rats and predators are expert diggers who can dig underneath, so the bottom of the coop needs to be secure.

The main point to watch out for with poultry housing in general is security from predators. Make sure doors can be fully secured, that there are no weak points in the structure and that air vents are not big enough to allow access to the coop. Predators can be very clever when it comes to getting a meal and have even been known to dislodge slide doors or droppings trays, so make sure sliding openings have bolts or are easy to add a bolt to in order to protect your birds.

The shape of the coop − whether ark, house shape or even chicken shaped! − is purely a question of your personal taste and requirements. What matters is that the house is safe, secure, easy to clean and dry for the girls so what it looks like on the outside is for you not them.

The triangular ark style with an under run is very practical and makes good use of limited space. Although we don't think that the run area is sufficient to keep the birds in all the time, it is useful in that you can leave your hens to come down for breakfast when they want and you can release them into the garden or main run when you get up. Don't forget that they work on sun time. You might prefer to be in bed when summer dawn rises at 5 am. Ensure the run underneath provides enough height for your birds and that the run underneath is fox-proof.

Incidentally, light- or timer-operated pop-hole automatic openers can be fitted to most wooden houses and are a real boon to poultry keepers who like their sleep. Pop-hole is just the name we give to the door the hens use to get in and out of the coop.

Lots of people choose pre-made coops and then paint them in fun, bright colours or, to hide them into the garden, in muted greens and browns. Even if you haven't the technical know-how to build your own coop, painting or staining your purchased coop with animal-safe paints makes it as individual as the chickens that will come to live in it.

Wooden coops will need maintenance, usually an annual treatment with preservative to prevent the wood from rotting. Always ensure you use a suitable preservative that is bird-friendly. In the UK where bats are a protected species, the preservative will usually state if it is suitable for bats. If it is OK for the bats, then it's safe for any poultry. If you are unsure, then contact the manufacturer who will be able to tell you.

One tip to preserve the wooden house is to put the legs onto bricks or small concrete slabs. Wood in contact with the earth tends to rot quite quickly. If the base sits on the ground, raising it on bricks will allow airflow under and prevent rot.

## Plastic Coops

The biggest innovation in hen coops for many years has to be the plastic coop. Some are now being manufactured from recycled plastic, adding to their green credentials. The benefits of the plastic coop are:

**Cleanliness** – Plastic houses have smooth surfaces which are easy to wash with a little washing up liquid and a soft brush before rinsing or just jet-wash clean.

**Pests and Parasites** – Unlike wooden coops, there are no nooks and crannies in the plastic to harbour pests like red mite. If you have red mite in the flock already, plastic is much easier to eradicate from.

**Low Maintenance** – Wooden hen houses need treating, usually each year, with bird-friendly wood preservative to keep them in good condition. Parts in contact with the ground are liable to rot and require repair. Plastic houses do not rot.

**Draughts** – Wood, being a natural material, will dry and shrink in hot, dry weather and expand in wet weather which can cause gaps to appear allowing draughts. Chickens are hardy but draughts can cause health problems, especially in the winter. A properly made plastic house should have no draughts or leaks.

**Long Life** – So long as the plastic is UV stabilized, a plastic house should never deteriorate and literally last decades.

**Insulation** – Plastic houses, especially in light colours, remain cool in the summer and warmer in the winter (lack of draughts), particularly if double skinned.

**Light Weight** – Plastic coops generally weigh less and so are easier to move around the garden or field.

**Environment** – Recycled plastic is helpful to the environment and reduces carbon emissions. Although wood for wooden houses can be re-grown, it depends on how the forest is actually managed as to how environmentally friendly it is.

**Appearance** – Many people love the modern look of the plastic houses.

There are some disadvantages to the plastic houses though.

**Cost** – It has to be said, they are a lot more expensive than cheap wooden houses to buy initially. However, when you divide the cost by the very long lifespan, plastic coops usually work out cheaper in the long run.

**Light Weight** – Although this is an advantage for moving the coop, in some very windy positions it could be a problem. In any case, chicken coops are best sited in sheltered spots.

**Appearance** – It's a matter of personal taste. Some people prefer the looks of the traditional wood hen coop or ark, and wooden hen houses can be painted and decorated to almost make them works of art!

**Fixed Design** – Wooden coops can be altered and extended if your flock increases but the plastic coop cannot be changed.

## The Eglu Range

The Eglu from Omlet was, we think, the first plastic poultry house to hit the market. The range is very much aimed at the backgarden keeper and smallholder and is ideal for those who only want to keep a few birds – no more than, say, ten.

The Eglu comes in a number of models which are all made from twin-walled plastic for thermal insulation. They have draught-free ventilation, keep cool in summer and warm in winter months. Perches and nest boxes are standard as are pull-out droppings trays. They also come in five colours and have many accessories, including runs, feeders and drinkers. The moulded and streamlined design appeals to many, although it is very much a love it or hate it thing. Owners tend to swear by their Eglu.

## Eco Hen Houses

The Eco Ark and Eco Hen Houses are award-winning for both their unique styling and for their green credentials due to the use of recycled plastic farm waste. They have adjustable ventilation to enable you to control the airflow and temperature depending on the season.

Currently we know of three styles, available in different sizes and colours. With the Eco Arks and some Eco Hen Houses you can buy integrated runs as well. Different models have different features but they all share the basic benefits of a plastic coop.

The main downside to all the plastic coops is the cost. At more than three times the price of a similar-sized wooden coop for the Eglu and double the price for the Eco Ark, this can be off-putting. We felt that you were paying a premium for the Eglu ranges, possibly to cover the cost of their marketing. However, as a long-term investment, the plastic coops make a lot of sense.

*Coops with Runs*
A number of coops come with integrated runs but often, although the coop itself can accommodate, say, four birds, the run only has enough space for half that number. When looking at integrated coops and runs, make sure that you follow the rule of at least 1 ft$^2$ (0.1 m$^2$) of sleep space and at least 1 m$^2$ (11 ft$^2$) of run space per chicken so that you don't suffer from problems with overcrowding and boredom in your flock.

For housing ducks, geese, turkeys and quail, see the relevant chapters.

**Building Your Own Coop**
The most economical quality housing is that which has been homemade. This also has the benefit that you can add exactly what you want to the house – multiple access points for eggs, high ceilings for cleaning, blockable nest boxes to stop the birds sleeping in them and soiling eggs, etc.

If your DIY skills are limited, then converting a shed or store may be the best starting point. With a small 6 ft x 4 ft (1.8 m x 1.2 m) shed, as well as a couple of lengths of timber for perches, you can make a coop big enough for up to 24 birds. Nest boxes can either be built from timber or you could convert old plastic boxes, wooden bedside cabinets, etc. As long as you put in plenty of bedding material, the chickens will be happy to lay in them. Make sure there is ventilation above head height and a pop-hole in the door that can be securely locked to prevent predator access at night.

If your shed has windows, try to site it so they are to the north. Overheating is as much of a danger if not more so than cold for your hens. If your windows catch the sun, then painting them white will allow light through but provide shade on warm days.

The economics are such that it can often be cheaper to buy a new shed to convert rather than build one from scratch. The manufacturer gets a better price than you can on the timber. Avoid the type constructed from overlapping timbers. They are draughty, almost impossible to clean thoroughly and will harbour mites. You want the type constructed from tongue and grooved boards.

Building a small chicken house from new is not that difficult for those with some woodworking skills. Rather than using tongue and groove, external quality plywood sheeting is quite easy to work with and the smooth surface makes cleaning and pest control easier.

Providing a removable droppings board makes sense, especially under the perching area where the majority of droppings will accumulate. Removable perches are essential for easy cleaning.

The nest boxes should be sited lower than the perch if possible. This discourages the hens from sleeping in the nest boxes.

A single-sheet, removable sloping roof made from corrugated roofing material such as Onduline with a good overhang helps to keep the walls dry and allows easy access for cleaning once again. The gaps aid ventilation, as well as the front and back air vents above head height. To prevent rats gaining entrance, cover gaps with weldmesh.

A sliding door enables the pop-hole to be closed and locked if required at night and a ramp with frets like a guitar enables the hens to climb up to the house if it is raised high to allow a run area underneath.

## POULTRY RUNS

Building a secure run for your poultry is a good idea, especially if you know there are predators in the area. Even if you don't think there are predators, you may be surprised by a ranging fox or even the neighbour's cat or dog appearing in your garden and getting to your birds unexpectedly. Foxes are a problem in both urban and rural areas and in some places you also need to consider stoats, weasels, mink and don't forget the two-legged predator, humans!

Runs do not need to be permanent structures. There are some excellent movable run options, as well as electric fencing that should be considered. As mentioned in the Housing section, many chicken coops come with integral runs. However, these are rarely suitable for long-term use for your poultry due to their restrictive size. Alternatives should be considered and planned before you bring your new birds home.

Larger poultry such as geese, turkeys and large breed ducks ideally should be either free-range with electric fencing or be kept in large movable structures. If you have to keep them in a permanent run, stocking densities should be low so that they remain happy in their living environment. Always take into account the space taken up by the housing, water pools (for ducks) and feeding area before calculating your available run space per bird.

When constructing a poultry run, you need to make sure you use a mesh that is fox-proof. Standard "chicken wire" doesn't have the strength to prevent a fox biting through the wire as it is usually only 19swg (1 mm thick). It is necessary to use a weld mesh of at least 16swg (1.6 mm) but ideally use either 14swg (2.05 mm) or 12swg (2.5 mm). Using a stronger mesh will also give support to the run structure.

When building a run, remember that foxes and other predators can dig as well as climb, so digging the mesh down by at least 18 inches (45 cm) or creating an 18 inch (45 cm) mesh skirting will stop them digging under. A movable run would obviously be better with the skirting and a fixed run should ideally be dug down.

To keep the fox out, the walls need to be 6 feet (1.8 metres) high and either have an outward facing angled overhang of 18 inches (45 cm) or the run will need roofing with wire to stop the fox gaining entrance. You can use ordinary and cheap chicken wire for the roof.

The benefit of electric fencing kits is that you can regularly move the birds to new pastures with ease, allowing the ground to recover. There are a lot of misconceptions about electric fences so an explanation is in order.

Basically the electric fence consists of a battery, an energizer, fencing wire, netting, posts, earth stake, insulators, fence tester and, if needed, a gate and gate handle.

The energizer takes a low voltage power source such as a battery and converts it into a high voltage current which is sent out in a pulse every second or two. Touching the fence causes the electricity to jump to the earth via whatever is in contact with it. This causes muscles to contract and is quite painful.

For those of us without an understanding of physics, the best way to explain it is to think of electricity as water. If you had a large barrel of water sitting on the ground and attached a hosepipe, the water would drain out but without much force. That's low voltage electricity.

Now if you put the barrel up high, the water would come out with more force and you could spray it for quite a distance. That's high pressure or voltage.

Finally, if instead of a hosepipe you had just a pinhole in the hosepipe, then the water (high voltage electricity) would squirt for a distance but there wouldn't be very much of it.

That's basically how the fence works: a very small amount of electricity but under pressure so it jumps a distance. Touching the fence is more painful to us than it is to a hen or a fox because they have the benefit of feathers or fur to act as an electrical insulator.

While it is painful, it's not a dangerous shock such as you could get from the household electric supply. The force behind the electricity is so small it's supplied by torch batteries in small fences. Larger fences may be fed from a car battery and these can be charged up from small solar panels with some kits.

Your poultry will learn very quickly (one or two shocks at most) not to touch the wire. Most foxes will leave the area after the first shock thinking something is attacking them and they'd best find easier prey elsewhere.

The three main things to remember when using electric fencing are:

1. Make sure the area isn't overgrown as grass and weeds touching the wire may short the fencing.

2. Regularly check the fence, energizer, insulators and earthing stake especially in wet weather.

3. Do not use with any additional form of barbed wire fencing as your poultry may get caught up in it.

If you choose to have a permanent run, unless it is quite large, you will find that the lush grass can be quickly turned into mud, especially when keeping waterfowl. There are many things that can be used as run bases once this happens including wood chippings, landscape bark, straw, pea shingle, rubber chippings and new turf.

Wood chippings and straw are cheap, but will need replacing regularly to keep the run fresh. Putting a garden mesh under the run cover will help it to last longer as it will prevent it sinking into the mud below and makes it easier to clean up when changing the base.

Rubber chippings and pea shingle are washable and although initially more expensive they can be disinfected and reused so may prove more economical long term. The main problem with pea shingle is that it can be uncomfortable for the poultry if it is the only flooring texture available to them.

Another option would be to regularly replace the turf within the run. This will keep the poultry happy but could prove to be expensive as, particularly in winter, they will decimate the vegetation if in a fixed area.

If possible using a moving run is the best choice as it allows access to fresh vegetation and the birds can be moved before an area becomes over used or damaged.

## BASIC EQUIPMENT

Obviously your new poultry are going to need to eat and drink. Making sure they have the right-size feeders and drinkers is important. A feeder or drinker can cost as little as £2.50 for a plastic one or over £25 for a heavy-weight galvanized one.

**Feeders**

Basic plastic feeders are great for those keeping a small number of chickens in the back garden. The only problem is that it is difficult to find a plastic feeder with a cover cap to keep the rain off. If you site your feeder in a dry place, perhaps under a raised ark, then that's not a problem.

Galvanized feeders are longer lasting and usually come with a rain cover, but they are more expensive to buy in the first place. We have found that for a small flock of three ex-barn hens, a feeder that holds 1.5 kg is more than sufficient for them. If keeping a larger flock of birds, it would be wise to invest in a larger feeder.

Ducks and geese not only eat more than chickens but also tend to shovel up the food with their flat beaks rather than peck food like the hen. From our experience, a galvanized feeder with a large open bottom feed area is easier for our 8 ducks to use.

Turkeys will also eat more feed so a larger feeder would be recommended for them too. Quail, on the other hand, like to scratch their feed and a small feed tray for them is ideal. We just use an old plastic takeaway container which seems to enable them to scratch and eat happily. We also use a shallow tray for hatchlings for the first week or two as they find it easier to eat from.

There are also lots of ways to make feeders from recycling. A length of plastic water guttering, especially the square section style, is great for ducks and geese. Just make sure that the feeder is covered to avoid feed spoilage. An old bucket can make a great wheat bucket for ducks (wheat with water filled over it). Old plastic containers, such as those used for our quail, can also be used to give treats to your flock.

If you have a larger flock and a problem with rats, mice or wild birds taking feed during the day, a treadle feeder may be ideal. These are large galvanized feeders with a tread step that opens a flap to reveal the food. Chickens, ducks, geese and turkeys quickly work out how these operate, although lighter bantams and chicks will not have the weight to operate one so they are better suited to the larger breeds of chickens.

**Feed Stores**

Rats and mice will be attracted to any feed that is left around in the evening after your birds have been tucked up for the night. Any feed left in the feeder should be taken out of the area and put somewhere secure for the night.

Feed bins should not be made of wood or plastic as a rat or mouse can easily chew through this and get at the food within, spoiling it. A galvanized bin (the old-fashioned dustbins are ideal) or larger metal feed store for those with larger flocks is recommended, with the feeder also being stored in it for the night.

One good tip we came across was to use a small dead chest freezer as a food store. Completely secure from rats and mice and easy to get it into, for a cost of nothing!

**Drinkers**

Drinkers, as with feeders, should be picked with your flock size in mind. A three-litre plastic drinker will suffice for a trio of chickens in the back garden whereas a larger flock or backgarden ducks and geese will need a larger drinker. One of the reasons we have always used plastic drinkers for chickens, even when we have had a larger flock, is due to the addition of Apple Cider Vinegar (see page 69) to the water that will damage a galvanized drinker.

Larger plastic drinkers can be obtained for bigger flocks and if you are intending to add water supplements we would suggest that you stick with these or have them in addition to a galvanized drinker.

There are some drawbacks with the plastic drinkers. The plastic gets scratched and quite brittle over time so they don't last as long as galvanized drinkers.

Ducks and geese will naturally splash in any water they are given. A galvanized bucket drinker is ideal for them as they can submerge their heads while drinking. You will also find that they drink a lot from the ponds and bathing water they are given. Again, old plastic buckets can be used. Just be warned they may try to get into the bucket for a swim!

## Bedding

Most poultry produce over 50 per cent of their droppings at night and so the poultry house quickly becomes mucky, especially with ducks and geese whose droppings are sloppier than hens.

There are many different types of bedding materials that are used for poultry housing. The four main ones are wood shavings, hemp bedding, shredded paper and straw:

**Wood Shavings** – A popular choice as they are cheap to buy and help to keep down the ammonia smell that is common in poultry houses. You should look for dust-extracted shavings as dust can otherwise cause respiratory problems within the poultry house. They provide a soft bedding and insulation for warmth in winter months.

**Hemp Bedding** – Often referred to under the two main trade names of Aubiose and Hemcore, this is derived from the inner core of the stem of the hemp plant. It is our favoured bedding choice, especially for ducks and geese, as it is super absorbent, comes with the option of added citronella which reduces flies, has a clean smell which reduces ammonia smells, is long lasting and, despite the initial outlay for a bale being more expensive than wood shavings, it will last twice the time so is cost-effective.

**Shredded Paper** – A cheap option for those with a home shredder. We have tried using shredded paper and will on occasion use it in nest boxes, but in the poultry house it quickly spoils and is unsuitable for ducks and geese as it needs refreshing on an almost daily basis. If keeping a small flock of chickens and you are able to regularly "poo pick" and refresh it, it provides a recyclable option.

**Straw** – This is something that can be used for layering nest boxes, but with heavier birds quickly becomes compacted in the poultry house itself. If using in the main house, chopped straw makes for a better bedding although we have found that it will need changing more often than wood shavings and hemp bedding materials.

However, straw is hollow so can keep moisture in and allow mould spores to grow. Dampness and the release of fungal spores are especially dangerous to chickens which are very susceptible to respiratory infections. For the same reason, hay should not be used in the poultry house. The hens may also eat the hay and develop an impacted crop as well. A good combination for most poultry keepers is to use wood shavings or hemp bedding in the main house and a layer of the shavings or hemp in the nest box with straw topping in the nesting area for comfort for laying.

**Cleaning Equipment**
The main equipment needed for cleaning out the poultry house includes a dust pan and brush, a cat litter scoop or a shovel if cleaning a larger house. The cat litter pan scoop is great for people with small flocks of chickens who can do a quick "poo pick" on a daily basis.

A dust pan and brush are needed when doing a full clean; a small shovel is required for a larger chicken house and especially for a duck house where the mess is often compacted.

**What to Do with the Waste**
You can, of course, pop the soiled bedding, etc., into bags in the dustbin but that's not exactly ecologically friendly and is wasteful. Most keen gardeners would be happy to take it off your hands; it's a fantastic fertilizer.

Don't use it immediately though. Poultry manure is too strong and can scorch your plants when used in quantity. The best thing is to add it to a compost bin as an activator. If you have one of those tall plastic compost bins that many councils give away now, pop in a layer of your poultry litter, followed by a layer of green material, grass cuttings or hedge trimmings.

A dusting of garden lime (just like the icing on a cake) on top of each layer of poultry litter is a good idea. It reduces the acidity, helping the microbes and worms do their job of changing raw manure to beneficial compost. It also stops smells and flies from being attracted.

In the winter, when you don't have much in the way of

green materials and grass clippings, add torn-up newspaper or brown (not glossy) cardboard and shredded paper. It will all rot down and in the summer you'll find that you have usable compost in just three or four months. If you can, mix the contents of your compost bin with a fork each month. This adds air and brings the undigested materials to the centre, thereby speeding up the process.

# 4

# CHOOSING AND OBTAINING YOUR POULTRY

When you start out it's all a bit confusing. There are so many different breeds, the list seems endless and the more you look at, the more confusing it becomes. Hopefully we can guide you through the maze so that the birds you choose will be right for you.

It's not our aim to list every breed available. Instead we're trying to cover the best choices for you to start out with. Once you've gained more experience, you'll be better equipped to decide what breeds are best for your circumstances.

Having chosen your breeds, the next problem is where to get them from. Contacting commercial suppliers isn't an option. They're not set up to sell fewer than a hundred or even a thousand chicks.

Livestock markets are another option but very much "buyer beware". If you have an experienced friend who can go with you, all well and good. There's no comeback if the birds you bought were not what you thought.

Except for rescue hens, it's best to look for a local breeder. Ideally go to a breeder who has been recommended to you by another poultry keeper. Failing that, there may be adverts in your local paper or look at the searchable list of breeders on our website. We must point out, however, that these are unchecked. Most small breeders raise poultry as a hobby but unscrupulous ones do exist.

Just like buying a car, if the seller is shifty and evasive, then the chances are that there is going to be something wrong. Take a look at the conditions the birds are being kept in: are they clean and healthy? You don't want to buy in health problems. We've come across breeders saying that you need a cockerel to keep your hens healthy, which is just their way of unloading excess stock. If in doubt, walk away and try someone else.

The ideal breeder will happily show you all his birds and how they're kept. In fact, he'll probably insist on showing them to you. You'll sense his pride and love for his hobby.

Before deciding on a breed, there are a few questions you need to consider.

*First: how much space do you have for your birds as this will restrict your numbers?* Although 6 bantam hens may be happy in a run that is 6 m$^2$ (64 ft$^2$), if you put 6 large hens in the same space, you would probably experience feather pecking and other problems as they would be overcrowded. This, along with making sure they have secure housing before you get your new birds, is the most important consideration.

There are usually no general legal restrictions on keeping poultry in your back garden but some properties may be subject to a restriction in the deeds or a local bylaw. Often restrictions in the deeds were written many years ago and are likely not to be a problem in reality. If you are uncertain or worried about a specific restriction in your deeds, it may be a good idea to contact your local Citizens Advice Bureau for advice on whether or not the restriction is still enforceable.

What you must not do is to allow your poultry to be a nuisance to your neighbours. The gentle cackle of a few hens or quacking of a couple of ducks is not a problem but a cockerel crowing at dawn or the volume from a lot of birds could well be. This could lead to disputes with the neighbours and visits from the local council and even abatement orders.

Noise is not the only potential problem. Ensure hygiene standards are high and waste is handled properly or you could attract rats and create a stink which would cause a nuisance.

If you have a large garden and keep a flock of over 50 birds you must, by law, register the flock with DEFRA.

If you're really stuck for space or live in a flat, then you may be able to keep your birds on an allotment. As a general rule, allotments should allow the keeping of hens and traditionally it was the norm. The problem is that keeping hens on the allotment almost died out and many societies and councils are now against the idea, which they see as change rather than returning to their roots.

Even if your allotment committee or council says no, it's not the end. Hens are permitted as of right (Section 12 of the Allotments Act 1950) and local bylaws cannot overturn Acts of Parliament. You can often get the committee to change its mind. When faced with a refusal from our council we downloaded a list of all our councillors' names and addresses and wrote to them individually pleading our cause and won. This excerpt from our letter may be useful if you find yourself writing to councillors, etc.

*I can see no problem with us keeping a few hens on our plots so long as they do not cause a nuisance and this would seem to be in keeping with the view of the House of Commons Committee for Environment, Transport and Regional Affairs in their report, as well as common practice on allotment sites nationally and in accordance with Section 12 of the Allotments Act 1950.*

*As you are probably aware, allotments provide both practical value to plot holders and therapeutic value. They provide access to affordable, good quality food, physical exercise, social interaction and promote mental health. Allotments also have significant benefits to the environment. I quote from the above report:*

*"Localized food production also brings environmental benefits by reducing the use of energy and materials for processing, packaging and distributing food: around 12 per cent of the nation's fuel consumption is spent on these activities. Many allotment sites also provide good examples of the principles of sustainable waste management with extensive re-use, recycling and composting taking place. Indeed, food growing in urban*

*areas has been recognized as an important component of sustainable development by the Food and Agriculture Organisation and the United Nations Development Programme."*

However, if you're going to keep your birds away from your home on an allotment, do think about security. In some areas allotments suffer badly from vandals and on one site they set fire to a chicken coop, killing the birds in the process. Theft is sadly a common problem as well.

*The second question is: what do you want from your birds?* Are you primarily interested in their egg-laying capacity, raising them for the table or are you more interested in their looks? With ducks, for example, if you want a good layer you would be best suited to the Campbell but for looks the Blue Swedish or Appleyard are lovely-looking with attractive markings and colouring.

*Third: are you now fully prepared?* Have you got the house built, run or electric netting secured, equipment cleaned and food at the ready? Once you have all of this set up, you can go and get your new birds. We're amazed how many people come home from a show with a few hens in a box and then wonder where to house them. Make sure you've covered all the points in the Preparing for Your Poultry chapter.

*Finally: what age are the birds you are seeking to buy?* Commercial producers of table birds will buy day-old chicks in volume. The remnants of the yolk sac will keep them going for a day or two so they're easy to transport across the country.

Some small-scale breeders will supply the hobby keeper with day-old birds as well. The cost is less than buying an older bird but you need to keep them under heat and, with laying hens, keep them for the 18 weeks or so until they start to pay their way.

Except for breeds where the sex is obvious at hatch, you'll be getting both cockerels and hens. This raises the problem of what to do with the males. You should also keep in mind that losses between chick and adult are going to happen and this can be very distressing to children.

Growers are birds from about 6 weeks old that are coming off chick crumb onto growers pellets. They require much less attention and special care than chicks, sexing is likely to be much more accurate and there will be fewer losses as they grow to maturity than with chicks. You still have to keep them for three months until they start to lay though.

Point of Lay birds are, as the name suggests, those nearly ready to start laying. They will be around 16–18 weeks old and you can expect them to start laying within a month to 6 weeks. For the backgarden keeper, this is the best age to get your hens. It gives them time to settle in and you to get to grips with the routines before laying starts. Do, however, check what your breeder means by "point of lay". Sometimes birds as young as 12 weeks are supplied who will, therefore, take longer to start laying.

## CHICKENS

One general and important point on chickens is that you don't need a cockerel. They make too much noise to be kept in a suburban garden. It's well nigh impossible to stop them crowing and it usually ends up with neighbours complaining and even noise abatement orders being served.

Even if you have very understanding neighbours who don't mind cock-a-doodle-do at 4 am on a summer morning, there is a problem if you try and keep more than one cockerel. They will probably fight each other and sometimes they will fight to the death.

The hens will lay the same number of eggs regardless of whether there is a cockerel or not. Obviously, without a cockerel the eggs will be infertile. The only reason to keep a cockerel is for breeding, although they do keep the ladies in the flock in order. Well in theory they keep order but nothing is certain in nature!

If you're thinking of buying chicks or young birds where you don't know the sex with the aim of keeping the hens for eggs and cocks for meat, do go for a multi-purpose pure breed as the conformation of hybrid birds for eggs means there won't be much meat on them. Also, do make sure you can cope with killing them when the time comes. Not being able to kill your bird is nothing to be ashamed of, but you do need to know beforehand.

**Rescued Hens**
Rescuing an ex-battery, ex-barn or ex-free-range hen can be one of the most rewarding experiences, but be warned it also comes with its own fair share of heartbreak.

Most rescue hens are 18 months old when taken from the sheds as they are deemed to be beyond their egg-laying prime. That isn't to say they will no longer lay, it just means that they are no longer commercially viable as they won't lay for six out of seven days a week. Although sometimes the hen may never lay another egg they often will lay three or four a week for at least another year after they have been rescued. Rescue hens should be seen as pets and most rescue centres will get you to sign an agreement that the hens are being kept as pets and will be treated as such.

Watching a rescue hen take its first few steps on grass, picking for the first worms or taking a dust bath under the tree is amazing and they very quickly become "proper" hens chasing around the garden and enjoying their second life.

There is, however, a downside to rescuing hens. Health problems due to the intensive way in which they were kept, such as extreme feather loss and prolapse, are common. Sometimes just the stress of the change can cause them to suffer from sudden onset heart attacks. Be prepared with the details of a good local poultry vet to contact in the event of veterinary treatment being required.

You should be aware that, when they first arrive, ex-battery hens can look pretty awful if they've lost most of their feathers. They will grow back but it takes time. Sometimes they just can't take the strain of the change in their life and die. If you have young children, they may be very upset by this, so it's best to prepare them. Plate 1 shows some ex-battery hens on their way to recovery.

Rescued hens have been known to live for up to four years after they are taken to new homes, but on average they will live another 12–24 months. There are many rescue centres that can be contacted but the major rehomers to contact are the British Hen Welfare Trust (www.bhwt.org.uk), Fresh Start for Hens (www.freshstartforhens.co.uk) and Little Hen Rescue (www.littlehenrescue.co.uk).

Another way of re-homing is to contact your local farmers directly and find out when they are next sending their hens for dispatch and arrange to collect directly from them. The re-homing fee ranges from 50p to £5 depending on the organization, although some farmers when you are collecting directly will give them to you for free so it is worth checking which is best locally for you.

When you first bring your rescue girls home, let them take life at their own pace. Make sure that if they have wobbly legs that they can get in and out of the coop safely. If they have feather loss and you take them in the winter, you can buy or make hen jumpers to keep them warm until they feather up again in the spring. However, if they wear jumpers, their ability to fly and escape predators is restricted so they must be kept in a secure run. Give them places to shelter and put food and water close to the coop so they don't have to go far until they have built up their strength and courage to explore their new home.

Rescue hens are by far the most rewarding poultry, but you should only rescue if you are sure you can cope with the problems that are likely to come with them and are willing to keep as a pet a chicken that may never lay again, but that will in time become a cheeky, happy and loved member of the family.

## Hybrids

Hybrid chickens are bred by combining two separate genetic lines for a specific commercial purpose, either for eggs or for meat. Here we're talking about those best suited for high yield egg production. Most hybrids will live until 3 years old with egg-laying tapering off at 18 months and usually ceasing by 30 months. Some hybrids can live longer and lay longer than this, but if it is a longer life and egg-laying span you are looking for then the pure breeds covered later will be better suited to you. Hybrids are often an ideal "starter" bird as they lay well and are easy to look after and are cheap to buy when compared with pure breeds.

The following are just some of the hybrid chickens you can buy. A lot of farms and poultry suppliers have their own names for their birds so don't be surprised to find a Warren referred to as an Isa Brown (in fact, this was the Warren's original name) or at a farm called Little Farmtown they may be referred to as Farmtown Browns.

**Warrens** (also known as Isa Browns, Goldlines and Bovan's Browns) – are the hens with the typical brown-hen looks which lay large, brown, well-formed eggs. They lay 320 eggs a year for their first laying year and 280 a year in their second then taper off quickly. Most rescue hens are Warrens as their temperament is calm and docile, leading to their suitability for the battery cage. In the back garden these are friendly birds that will come running when they see you and will learn quickly when it is treat time! They are an ideal first bird due to their lovely temperament.

**Speckledy Hens** – a cross created from Marans and Rhode Island Reds, these are pretty birds with a similar colouring to their Maran parent but with the laying capability of around 320 eggs a year for their first laying year and 280 a year in their second then as with the Warren tapering off quickly. Their eggs are large and dark brown in colour. They are very docile and easy to handle so well suited to families with children.

**Amber Stars** (also known as Amberlinks) – are light-coloured birds that will lay up to 320 eggs a year for their first laying year and 280 a year in their second then again tapering off quickly after this. The eggs are light brown in colour and medium to large in size. Some owners have found them to be docile, friendly birds that are good with children and easily handled. However, some people report that they can be less docile with other birds and are fond of feather pecking which means they are not suitable for smaller pens and are better suited to free-range conditions. They can be flighty until they start laying and have been known to clear a 6 foot (1.8 m) fence.

**White Stars** – a hybrid that lays medium to large white eggs with 300 eggs a year for the first laying year and 260 a year in the second then again tapering off quickly after this. They can be flighty and often need to have one wing clipped if keeping them in an enclosed garden.

**Skylines** (also known as Columbines) – these are hybrids with Cream Legbar parentage and because of that they usually lay a large blue egg. They will lay up to 320 eggs a year for their first laying year and 280 a year in their second then again tapering off quickly after this. Again, they can be a little on the flighty side and wing clipping is advisable to stop them escaping into next door's vegetable patch!

**Black Rocks** – these come from Rhode Island Red and Barred Plymouth Rock parentage and unusually for hybrids are long lived often reaching 8 years. They produce between 230–280 brown eggs a year and continue laying for five years on average. They're a very hardy breed ideally suited for free-ranging.

**Pure Breeds**
For both looks and long-term egg production, you should look at Pure Breeds, some of which come in both large and bantam breeds. Large birds obviously lay larger eggs than the smaller bantams but, if space is at a premium, a fresh small egg is better than overcrowding.

Although Pure Breeds are more prone to broodiness than hybrids anyway, the bantams are often even more broody so this is something else you need to prepare for. We discuss this further in the Breeding Your Own chapter.

When it comes to bantams you should also not assume they are quieter birds as bantam cockerels can crow just as loud, if not louder, than their large fowl counterparts! They are better suited to the smaller garden or run, but come summer and winter time they often need more protection from the elements so rain and sun shelters are a must.

Most of the birds listed below come in both large fowl and bantam size.

**Sussex** – this breed comes in many colours including Light, White and Speckled, and in large and bantam size. The large fowl versions are a dual-purpose bird (i.e. good for both egg-laying and meat), with an 8 lb (3.6 kg) or so dressed weight for cockerels. They have a good docile nature and make great backgarden birds for people with children or pets as they are not easily scared. Their eggs are cream to light brown in colour and you can expect to get about 240–260 eggs a year with steady egg-laying until they are around 3 or 4 years old and then a gradual slowing down with age.

**Faverolles** – these are quiet and gentle birds that are ideal for children and although they come in many colours the Salmon Faverolle is the most popular coloration. Faverolles come in large and bantam sizes. The large fowl birds can be dual-purpose with the cockerels reaching a very reasonable table weight of up to 9 lb (4 kg). The girls should lay around 200 eggs a year until they are around 4 years old with a slow decline after this.

**Welsummer** – another attractive and docile bird that is ideal for backgarden poultry keepers and that comes in both large and bantam size. The hens are a lovely rusty red and orangey brown colour. The Welsummer hen should lay a lovely dark brown egg and although they tend only to lay around 180 large (70g/2.5 oz or so) eggs a year they will lay for around four years before gradually slowing down. Again the Welsummer cockerel can make a good-sized table bird with an 8 lb (3.6 kg) or so dressed weight.

**Leghorn** – the Leghorn can be white, black or brown in colouring. Both the hen and cockerel have wonderful fan tail feathers. The large fowl cockerels are too light to make a good table bird although if they can't be rehomed or kept they can be deboned – they would be too small a roasting bird but you should be able to get a small meal from them. The hens lay about 240–260 white eggs a year for the first three or so years and then slowly decline in numbers as they go into old age. They can be a little flighty and, if keeping in an enclosed area, wing clipping may be necessary.

**Araucana** – a very attractive bird with a very small comb and "pompom" crest, and can also be bred in a rumpless (tail less) variety as well as large fowl and bantam size. The Araucana comes in a number of colours but the lavender, blue-red and blue are the most popular. They lay lovely blue eggs but, depending on the purity of the breed strain, this can vary from olive green through to violet shades as well. They will lay up to 200 eggs a year and this will decline after their third or fourth year. As with the Leghorn, the large fowl cockerels are not good table birds but after deboning can make a small meal. The Araucana is naturally a broody breed and will make a good mother. She will often also take on incubated day-old chicks and raise them with her own.

**Barnevelder** – this is a breed developed for hardiness and as such is good in cold/wet/windy climates. It is another dual-purpose bird in its large fowl version. A large bird with a friendly, but somewhat lazy, reputation best suited to free-range environments. The cockerel should make a 6–8 lb (2.7–3.6 kg) dressed weight and the hen will lay up to 280 large brown eggs a year for the first three or four years. The Barnevelder comes in a number of colours with dark brown/black, double laced partridge and blue laced.

**Cream Legbar** – another blue/green egg-laying breed. Day-old chicks can be sexed by checking the colour of their feathers. This is known as autosexing. See Plate 2. It is beneficial to those unable to keep the cockerels as they don't make very good table birds. They are very flighty and although on occasion can be friendly they are not best suited to noisy gardens or children. The egg colour again depends on the purity of the breed strain, but you can expect around 260 blue-green eggs a year until they reach 3 years plus. Like the Araucana, from whom they were bred, they have a crest and small comb. Colour wise they have silver-grey main feathering and creamy-salmon breast feathering.

**Gold Legbar** – similar to the Cream Legbar in that it is an autosexing breed and with grey/brown feathering, a brown or golden neck hackle and a salmon breast. The Gold Legbar lays a white egg with around 220–260 a year in their first three years. They are a light breed so are not best as table birds. Wing clipping is advisable as they can be flighty with their light body weight.

**Maran** – a lovely large and docile layer of dark brown eggs. They are suited to larger back gardens and good with children as they can be very friendly and inquisitive. But they are prone to getting a little large and if they don't have enough room to explore are quite destructive, so a large run or free-range in the garden is best. They can be different colours but the most common is the speckledy grey-silver cuckoo colouring. They will lay around 220 eggs a year for the first three or four years and the large fowl cockerels, although slow growing, can get a table weight of 6–8 lb (2.7–3.6 kg) dressed.

**Naked Necks** – these are very much an acquired taste as far as looks go! The Naked Necks look like someone has crossed a turkey and a chicken, resulting in completely naked faces and necks. They are also heavy birds, making them ideal for someone breeding for eggs and meat as they get a good dressed weight of 6–8 lb (2.7–3.6 kg) and are easy to pluck and prepare due to the lack of feathering. Naked Necks can have red, black, cuckoo or blue feathering and they lay large brown eggs with about 220 a year for the first three years.

**Rhode Island Red** – a very traditional chicken breed with lovely red brown feathering and a friendly personality. Perfect for people with children or other pets. The large fowl counterparts make good dual-purpose birds with a dressed weight of about 8 lb (3.6 kg). They lay large brown eggs and will exceed

260 for their first few years and then slowly decline in numbers.

**Silkies** – these have a wonderful fluffy and silky plumage which accounts for their name. They have a black/blue skin and meat which is quite unusual in itself and they also have blue ear lobes and five toes. They really are the ideal pet, especially with children as they are so affectionate and friendly with their owners.

Even the large version of the Silkie is small and with their feathered feet they need protection from the elements and predator-proof surroundings. Silkies come in many colours, with the white, black and gold being the most popular. They can also come bearded or unbearded, as well as having their "pompom" hairdo.

As for eggs, they will lay around 160 small cream eggs a year but they do go broody easily so they may lay fewer. Due to their black meat and small size, it is uncommon for them to be used for meat, although in China and a few other countries their meat is favoured by chefs as a delicacy.

They tend to broodiness and make great mothers so, even if you are breeding other breeds, a Silkie or two is worth having as a surrogate mother to sit on the eggs and raise the chicks.

**Seabrights** – gold and silver Seabrights are small birds and a true bantam as they have no large fowl version. They are mainly kept for pets and showing and as such lay very few eggs – perhaps 80 a year. They are friendly, but given the chance they will roost, so care needs to be taken in urban gardens to stop them escaping.

Roosting is when the bird perches on branches or at night on the roosting bars in the chicken house. See Plate 4. When they roost, it stops them from sleeping in their own droppings as chickens will produce the majority of their droppings while sleeping.

Because of their size, they need good weather and predator protection.

# 5

# WHEN YOU GET
# YOUR HENS HOME

Apart from chicks who should be cared for as described in the Breeding Your Own chapter, your new birds are likely to be stressed from the trauma of being removed from the environment they are used to and the journey to their new home.

The temptation is to pop them down on the lawn and show them to the family who will all want to fuss with the new arrivals. This will add to the stress of your birds and isn't a good idea.

It's best to put them into their house with some food and water and shut them in for the night. This will calm them down and after a good night's sleep they will be rested and ready to face a new world.

They will have accepted the new house as home and that will make life easier as they will be less likely to try to find their way back where they came from or to roost in a tree come nightfall.

Open the pop-hole and let them come out in their own time. Most will come out immediately but with rescued hens it can take longer, even a couple of days, for them to summon up the courage or even realize that they can range further than the house. Placing food and water just outside the house or making a trail of treats will help encourage them out for the first time.

Sit quietly with your birds on your own and keep the rest of the family and especially dogs away until they're well settled. A few treats such as meal worms will help establish you as a friend and the provider of all good things.

As they'll have come from a larger flock, the chances are that there will be a little squabbling as they establish who is who and develop a new pecking order. Normally this isn't too much of a problem but keep an eye out for injuries inflicted in the squabble.

Once the birds have been encouraged out, you may find that getting them to bed the first few nights is difficult! We have always restricted treat time to bedtime as this encourages them back to the poultry house. As with encouraging them out of the house for the first time, encouraging them back in with treats such as mealworms will help to set them into a bedtime routine. In our experience, new birds will happily follow a rattling cup of corn back to their beds.

# 6

# INTRODUCING THE
# NEW BIRDS TO THE FLOCK

Chickens in particular are social animals. They can recognize and distinguish up to 50 other birds in the flock and they all know their place. If there is a cockerel around, he's normally the boss and helps to keep order, calling the flock to food and defending from predators.

Below him is a hierarchy, the famous pecking order, where the hens rank themselves into order of status. Without a cockerel, the hens will decide amongst themselves who is on the top and who is on the bottom of the social heap.

Once they've established the order, there's very little argument between the birds. Occasionally one bird will decide she can move up the ranks and there's a squabble but it's not a major issue. However, when you decide to increase the flock by adding new birds everything changes.

The tendency is for the existing flock to drive off strangers who will compete for food in their territory. The fights that develop if the introduction is not handled carefully can often result in injury and even, in extreme cases, death.

Bring home your new birds and, as with your first birds, keep them in a separate house and run overnight with food and water available. When you let them out the next day, keep them in the run which should be visible to the existing flock but do not allow physical contact between them.

The idea is for all the birds to be able to see each other but not to interact physically. You may need to attract your existing flock over by feeding some treats to both groups from outside the run.

Apart from the risk of fighting, keeping the new birds separate for two weeks quarantines them in case of illness and helps prevent introducing disease into the flock. Even if the new birds are from the same breeder as your first birds or look perfectly healthy, a disease may have broken out or be incubating unseen.

After a fortnight, your hens will all have become used to seeing each other and, with a little luck, consider themselves part of the same flock. Release the new ones and they should integrate into the main flock without too much discussion. Be on hand to intervene if a serious fight should break out though. In the evening, put them into the same house as the existing flock and hopefully that should be that.

If a hen is injured in the process of being introduced to a new flock or in a "dominance discussion" in the existing flock, then she should be treated and separated for a day or two in the separate house and run until she's fully healed. Ideally this run should be in visual contact with the rest of the flock.

Unfortunately, hens have a bit of a mob mentality and your bleeding hen may be attacked by a group. The way to handle this is to remove the hen until she is both feeling better and looking less like a target to the others.

Obviously those with a smaller back garden may not have the space to keep the birds separate. Ninety-nine per cent of the time the quarantine is unnecessary, so you'll most probably be OK on the disease front.

It's the pecking order front that is most likely to be a problem. You should try to keep a close eye on your existing flock and the new birds over the first week or so. Putting the new birds into the housing at night and making sure that when you first unlock them in the morning you are around for a while to intervene if birds become aggressive with each other is best.

Remember that some battling and pecking is natural with all poultry as they establish the "pecking order". However, if one bird is being particularly picked on, or if they draw blood, intervene.

If blood is drawn, you will need antiseptic spray, such as the purple equine spray, to clean the wound and cover the colour of the blood. Otherwise this area will be persistently attacked if the other birds can still see the blood. This mob behaviour at the sight of blood is a problem especially with chickens but also with quail.

## Introducing Mixed Flocks

If after, for example, keeping chickens for a little while you decide you would like to add a few ducks to your backgarden flock, you will need to make sure, as with same species introductions, that the introduction is carried out carefully.

Different species will require separate housing, but they can usually be kept together in the same run or garden with little difficulty. The main exception to this are turkeys, which are covered in their own chapter. You will tend to find that introducing ducks to an existing flock of chickens is fairly easy and most of the time they will ignore one another apart from trying to steal the others' food! Watching over the birds for the first few days is a must, however, just in case trouble breaks out.

Most chickens are safe around a pond that is kept for ducks or geese. Apart from occasionally choosing it to drink from, they will ignore the pond area. Providing separate drinkers and feeders for the species is a good idea though.

Ducks and geese will make all water mucky by the end of the day and the chickens should have fresh water available, although they often choose to drink from the muddied water instead!

Having a few separate feeders around is recommended as with same species flocks to prevent any bullying or arguments. Whilst chickens have a sharp beak (great for pecking with), the waterfowl have a bill which is better adapted for shovelling so it's best to provide a trough for the ducks or geese.

Be aware that medicated feed with an anticoccidial for chicks is positively harmful for your ducks. Even if you supply separate feeds, the ducks are bound to go for the chicken feed and vice versa.

We have noticed that our chickens are the bullies, not the ducks. The chickens will peck at the ducks until they have eaten what they want and are very clearly the bosses of the garden.

Ducks and geese make more mess than chickens and come winter, although the ducks may not mind the mud, the chickens will be unhappy if wading to their ankles in a muddied run or garden. If you have limited space, try to keep them separate during the wetter months. If your garden is large, this should not be a problem provided you do not over populate your birds.

# 7

# FEEDING AND WATERING

## FEEDING CHICKENS

In the wild, all our poultry would find its own food. Chickens
naturally live on quite a mixed omnivorous diet. They'll eat
some greens, some grains and scratch about to expose worms
and other insects. Surprisingly they are partial to some meat
and will eat carrion. If you have ever seen them go for a
mouse, you'll be under no illusions. Beneath that cuddly,
feathered exterior there's a velociraptor straight from the film
*Jurassic Park*!

You might think from that you can just let your birds roam
around the garden feeding themselves with, perhaps, the odd
handful of grain scattered on the ground once a day.

The facts are somewhat different. Many of our chickens
could possibly just about keep going in the wild. There's even
a large roundabout in Norfolk that was famous for its flock of
wild chickens. Despite them being said to be wild, it's more
probable that they were actually being fed. When there's
publicity to be gained, trust nothing!

We've selectively bred our chickens, both pure breeds and
hybrids, to be far more productive than their wild ancestors.
No wild bird could eat enough to lay 200 or more eggs a year.
To achieve this they need to take in the right amount of nutrition to convert to all those eggs as well as keep themselves
healthy.

Humans react to higher demands for food by eating more. We have an advantage in our digestive system: we chew up our food and pass it to the stomach which then passes it to the intestines. As all of us who are getting a little older know, stomachs and waistlines can expand to take on more supplies.

Birds have a different system. They don't have teeth for a start and so can't chew their food. They tear off bite-sized chunks or swallow whole and this goes into their crop, which is a sort of holding area. You can feel when a hen has eaten well as the crop in her breast will be swollen. From there the food is passed into the small glandular stomach called the proventriculus. This moistens the food, and the secretions start the process of breaking it down. From there it goes into the gizzard. The gizzard is a cavity surrounded by muscles which contains some small stones and grit. The muscles contract and squeeze, which causes the stones to grind up the food, the equivalent of our teeth but in the wrong order. From there it goes through the intestines and, once the goodness has been extracted, comes out of the other end as useful fertilizer.

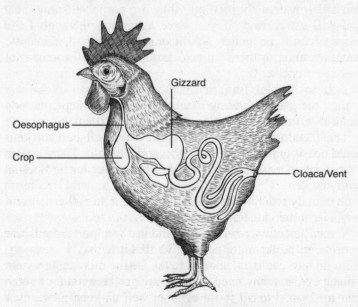

**Fig. 1**    The Chicken's Digestive System

The problem with this system is that the capacity is limited. Once your chicken's crop is full, that is that. No more can go in until that food is digested. If you allow your birds to fill up on treats and greens then they just won't have the room to take onboard food containing the nutrients they need. They're just like children, really. They'll fill up on sweets but have no room for dinner.

The best and easiest way to feed your birds is with bought-in compound feed. A compound feed is simply a feed that has been manufactured to include all the necessary vitamins and minerals along with a balanced amount of protein, oil, fibre and carbohydrates.

For adult birds the compound feed comes in two types: pellets or mash, sometimes called meal. In our experience, the pellets are far easier with no messing about; you just pour into the feeder.

There are various types of feed available, with mixes formulated for free-ranging birds, layers, etc. A basic layers pellet feed suits our birds. You can also get organic feeds, for a hefty supplement.

Do be aware that hens can be very fussy. Changing the brand of feed can cause them to stop eating for a while. We moved from a cheap feed to a dearer one as the cheap feed was out of stock. Would they eat it? You'd have thought we were trying to poison them.

For chicks with their specific nutritional requirements, you buy a specific chick crumb, which is easy for them to peck and is formulated to help them grow. You can also buy feeds with added medication. It's almost the norm to find chick crumb containing coccidiostats (see the Diseases and Problems chapter).

Once they reach 6–8 weeks old, it's time to move them on to growers pellets. To help them make the transition (as we said, they're fussy creatures), for the last week on crumb, mix the growers pellets into the food 50/50. Once they're grown to point of lay around 18 weeks old, again mix the growers pellets 50/50 with layers pellets to get them used to the change. Don't worry if the mother hen also eats the chick crumbs, they won't do her any harm.

Although pellets are easier to use, if you take on new birds they may prefer the mash and not even recognize the pellets as food. Rescued hens, who may have been subjected to beak trimming, will find mash easier to cope with.

As a general rule, the first meal of the day should be compound feed. Their crops will be empty and so they'll fill up on the healthy option. Do make sure the feeders are full and available all day so they can pick ad lib. Remember, in the wild, eating is what they do all day and their system is geared to little and often.

Other feed types such as wheat and corn should be fed as treats only in moderation and after the chickens have eaten a good quantity of their layers pellets for the day. It's best to give them these supplementary grains by scattering a small handful on the ground. This can also be a good way of getting your chickens back into the poultry house at night; you will find that, after only a few days, your girls will sit waiting at "treat" time. So it also forms a good bonding with your pets and can help tame even the flightiest bird.

If your birds are kept in a run and they no longer have grass or greenery to peck at, it may be advisable to add greens to their diet both to avoid boredom and to improve their diet. Cabbages on sticks or hung from the run's roof go down particularly well.

Poultry make good use of garden waste that would otherwise end up in the compost bin. A sweetcorn harvested a little late with tough kernels, bolted lettuce, cabbage leaves nibbled by slugs or caterpillars and wilted comfrey leaves are all very acceptable both as a treat and a supplement. But their favourite has to be tomatoes. We don't know why but they absolutely love tomatoes. Never let them in the greenhouse!

To quote that great exponent of self-sufficiency, John Seymour: "You can make the best compost in the world in twelve hours by putting vegetable matter through the guts of an animal."

Another favourite of our poultry is cooked pasta, especially in a tomato sauce, served warm on a cold day. We've come across people giving their poultry warm porridge for breakfast on a cold day. The hens may like it, but perhaps it's better served in the afternoon after they've had their layers pellets.

You shouldn't give your birds plate scrapings or any waste that has been in contact with meat. Outer leaves cut off a cabbage in the kitchen are fine but waste you'd normally put in the bin is not.

Some people will grow grass in trays and place this in the run, giving the birds something to scratch at as well as eat. Birds that are elderly, ex-battery or generally more infirm can have their greens chopped and shredded and spread on the ground to scratch and peck at.

We mentioned that they eat insects and meat. If you suffer from slugs and snails in your garden, try laying small pieces of wood down on the ground, lifted about half an inch (1 cm) with a stone or two at one end. In the morning flip the wood over exposing the slugs hiding in there for the day and watch your birds gobble them up. You can pick off the slugs and snails and throw them into the run as well. Much better than throwing them into the neighbour's garden from which they'll probably crawl back.

Incidentally, we'd advise against using metaldehyde slug pellets in the area of the garden where the chickens roam. The manufacturers have made the pellets unpalatable to birds and pets so the risk of them directly eating the pellets is small but if your hens find a poisoned slug they will eat that. We're not saying they would definitely become ill, but we're not going to experiment and find out either.

The ferrous phosphate pellets are much safer but if you have a major slug problem in the garden then we'd let the hens deal with them directly. They prefer slugs to vegetables so will deal with them first and then you can put them out of the forbidden area. Incidentally, ducks are even more mad for slugs than chickens.

If you dig over your garden border or vegetable plot, let the poultry range over it. They'll not only remove the slugs but also the slug's eggs and many other pests and weed seeds. Poultry are great for cleaning the ground.

We've discovered that poultry are very fond of cat food too. This probably applies to dog food as well. Leaving the kitchen door open in the summer, we found the hens squabbling over the cat's food bowl. The cat sat there watching, looking very sorry for himself. Sometimes a little cat food can be useful to perk up an off-condition bird but not too much or too often.

Another treat for your birds are worms. If you have a wormery, then you have a treat-making machine for chickens. Throw down some compost containing worms and they'll clean them out in no time flat. You can also buy meal worms to give your birds. They love those as well.

You can buy your poultry feed from a number of places now. Many of the chain pet stores have started selling poultry feed along with other supplies, albeit at fairly high prices. You can obtain your feed online but the cost of delivery can bump up the price.

The best places we have found for poultry feed and supplies are local farmers' feed supply shops. As of March 2011, for example, a 20 kg (44 lb) bag of layers pellets from a large pet store in Leeds cost £10.50 whereas the same 20 kg bag cost £7.25 at a local feed farm store and a locally produced brand of layers pellets cost £7.50 for 25 kg (55 lb).

Usually layers pellets will be in 20 kg or 25 kg bags but you may find chick crumb and growers pellets available in smaller bags. So long as it is kept dry in a vermin-proof store such as a galvanized bin, it will keep in good condition. Check with your supplier or the instructions on the bag if in doubt about its storage life; vitamins can break down over time reducing the efficacy of the feed.

Wheat and corn are more readily available in larger bags from the larger feed and farm stores, whereas the pet stores tend to stock 5 kg (11 lb) bags that cost almost as much as a 25 kg (55 lb) sack from a larger supplier. It is worth asking locally to find your best supplier before you start to keep poultry. Other poultry keepers will help and, if you cannot find any locally, check online for suppliers and advice.

Even if you are unable to sell eggs as organic, some people choose to feed their flock organically for taste or environmental reasons. On average, organic feed is one third more expensive than conventional feeds and you will find it harder to source than conventional feed.

The feed you choose is a personal choice, but remember

that as with any animal if you are changing your flock's diet you should do so gradually to avoid illness or a refusal to eat the new feed.

### Grit and Oyster Shell

Whilst, in theory, compound feed includes all the required nutrients including grit and calcium, you should provide some additional chicken grit in case they can't pick up enough small stones for their gizzards. Flint grit is said to be the best and costs very little, around £1.40 for a kilogram (2.2 lb) at the time of writing. Just put some in an old margarine tub or a pot and the chickens will peck at it as they need.

Crushed Oyster Shell provides a source of soluble calcium, which is essential for strong eggshells and bone formation. If your hens start laying soft-shelled eggs, the most likely cause is calcium deficiency that crushed oyster shell will remedy. Once again it is very cheap to buy and should be provided in the same way as grit, ad lib. The birds seem to know when they lack it and just take what they need.

You may have heard the recommendation to feed your hens crushed, roasted eggshells to replace the lost minerals. You can do this although it's a lot of fuss to save pennies and probably costs more in electricity than it saves anyway. If you do this, it is critical that the shells are really well crushed. The last thing you want to do is teach your birds to eat eggs.

If you have a problem with your birds knocking over the containers or fouling them, you can buy a grit block to hang in the run. This provides both the grit and oyster shell required, along with maize to attract the birds. Slightly more expensive but it lasts for ages anyway.

### Poultry Spice

This is a mineral supplement used by poultry keepers in a similar way to ACV to increase the condition and health of their flock. It has a high mineral content and is considered to be beneficial for birds that are a little drained, for example those who are laying soft-shelled eggs or going through heavy moults.

We will on occasion add some Poultry Spice to our hens'

diet if one seems unwell. Sometimes in the winter months, making a warm wet mash from their layers pellets and adding a teaspoon of spice seems to perk them up on the colder mornings. When we got our ex-barn girls, in addition to making sure they had lots of layers mash, ground oyster shell and grit, we also added Poultry Spice regularly for the first three months to help them refeather and thrive.

## FEEDING DUCKS, GEESE, TURKEYS AND QUAIL

Adult ducks, geese, turkeys and quail if kept with chickens can be fed on the same layers pellets. It is advisable when rearing young to use waterfowl or breed-specific chick crumb and growers before moving them to a layers diet.

Do be careful not to give your ducks, etc., a feed medicated with a coccidiostat which can be harmful to them. Check your bags carefully or ask your supplier, especially when buying chick crumb.

Geese enjoy grazing on grass. If you have enough land, they can be fed on a short grass area as they will eat the grass and the bugs, snails, slugs and worms they find while foraging during the summer months. If you're lucky enough to have a field, they make a great alternative to sheep.

However, most people will not have the extensive grass areas to do this and as such a supplemental layers or water-fowl feed will be required all year round.

A wheat bucket is ideal for those keeping mixed flocks of chickens and waterfowl as it provides wheat for the waterfowl while preventing the chickens from filling up on a feed that is less beneficial for their diet.

Get a bucket that is at least 1 ft (30 cm) deep and put a few scoops of wheat in there, topped up with water to over the fowl's head height. The waterfowl will enjoy eating the wheat as it enables them to sift it, as well as being more easily digestible due to the soaking; and it stops the chickens eating it as they would need to submerge their heads below water to get at it.

## WATERING

Many poultry keepers occasionally add a supplement, such as Apple Cider Vinegar (ACV), to the water they give their birds to help improve the general condition within their flock.

### Apple Cider Vinegar

Apple Cider Vinegar is a natural antiseptic with vitamins, minerals and trace elements that are all beneficial for the backgarden poultry flock. It also naturally lowers the pH within the stomach helping to kill off nasties as well as clearing the airways in chickens. Its many uses mean that poultry keepers often refer to it as a cure-all. This is obviously not the case but it is very much a good all-round store cupboard staple to keep when raising poultry.

For an adult flock of hens or other poultry (excluding quail), the ACV should be diluted roughly at 25 ml per litre of water. If using for younger birds or quail, dilute the mixture further to 5 ml per litre of water. As mentioned elsewhere, only plastic drinkers should be used because galvanized drinkers will corrode with the use of ACV.

When buying ACV, make sure you get unfiltered ACV and not the type from the supermarket which will have been pasteurized. You should be able to find it easily at a local equine feed store or where you purchase your poultry feed.

We give our hens ACV for one week in their water then they have three weeks without it added. With the ducks, we only give it to them when they have had a heavy moult or seem run down. However, many keepers will use it for all their flock on a regular basis. The way keepers give their ACV differs from person to person. Some add it two days a week, others five days a month and others the same as us: one week on, three weeks off. It doesn't need to be given every day and is just added when filling the water containers, so you can decide for yourself how often you wish to include it in their diet.

# 8

# CARING FOR
# YOUR POULTRY

*Losses are also often traceable to lack of oversight on the
farmer's part. Feeding may be irregular and unsuitable; food
troughs and drinking vessels may be foul; slight injuries and
disorders may go untended; and the stockmen are negligent
because the master is not observant and strict enough.*

From: *The Burden of Disease* by E R Ranson,
published in the *1924 Country Gentlemen's Estate Book*

There are two legs on which good husbandry of your flock
stands: routine and observation. Any pet owner will tell you
that animals are creatures of routine. Regardless of her
inability to read a clock, the cat will expect tea to be served at
the usual time and be waiting by the bowl come 5 o'clock or
whatever time she's accustomed to being fed. The dog will be
sat looking at his lead come morning walk time, despite being
no better at reading the clock than the cat. Poultry are just the
same, creatures of habit and routine.

Developing a routine and sticking with it is important to
your flock's well-being. Not just for their happiness but also
their physical condition. Feeding correctly and at the right
time ensures they get the nutrition they need. Once established, your routine will give you the time to manage and
observe the flock without too much thought on the details,
which you handle on auto-pilot.

Observation, the second leg, is critical. We don't mean just counting them in and out, but looking at the detail of their condition, their housing and fencing so problems are picked up before they become serious. There's a lot of truth in the phrase "a stitch in time saves nine".

## The Daily, Weekly and Monthly Routine
On a daily basis you will need to let out and lock up your birds. Obviously the amount of daylight changes depending on the season and poultry naturally get up with the dawn and head for bed when dusk falls. We tend to let the birds out an hour after it gets light during the winter months and in the summer months stick to 8 am every day.

Leaving the birds for an hour after dawn and putting them to bed before dusk means they are safely tucked up at the prime time for the fox to be hunting. However, you should never assume that you're safe from him at any time of day or night.

The automatic door openers are a real boon to those of us who enjoy the occasional lie-in. If you have a fox-proof run, then you can sleep a little longer knowing that the birds are let out and safe. If you are one of those people who leap up with the lark at the crack of dawn, then perhaps it's not so valuable.

Chickens, in particular, can be very noisy if kept cooped up or if their routine is altered. On weekends, lie-ins can be difficult if they start cluck-clucking loudly or if a cockerel decides it is time to "get up now, please!"

Fresh food and water should be put out every day and the grit and oyster shell containers checked and filled as required. Cast your eye over the birds and make sure they're all looking healthy and happy. Are they behaving normally or is one hanging back? Any signs of injury? Any bullying going on? Keep chatting to them all the time so they're used to your voice, the provider of all good things.

Take a minute to look over the house and run. Is everything secure? Any signs that a fox has been trying to dig under the wire or chew through? Check around the base of the house as well for gnawing – an indicator that rats are around. If you rely on an electric fence, check it is actually working and watch out for fallen branches or growing plants shorting it out.

Check the nest boxes as well. Leaving eggs around all day is a temptation for one of them to start egg-eating. What's the weather like? Make sure that feeders are covered during wet weather to avoid the feed spoiling and being wasted.

If you are about in the afternoon, then this is your chance for some quality time with your pets. Even on a cold winter's day, sitting and enjoying their company for a little while beats being indoors. Collect any eggs they may have laid and now's the time to give them a treat. It might just be some grain, or some meal worms or a tomato or two. On a warm day, double check the water supply. They can get by without food for a bit but lack of water is a killer. If you have some greens – maybe the outer leaves of a cabbage or a bolted lettuce – now is the best time to let them have it.

When it comes to bedtime, you will find that, as it approaches dusk, most birds will take themselves back to their house of their own accord. If they don't, using a treat trail or herding them for a few weeks should get them into this habit. Always make sure all pop-holes and doors are securely closed and it is a good idea, even if using an auto opener and closer, to double check at night that they are safe and secure.

Do not leave feeders out at night, as this will encourage pests such as rats and mice in the same way as spoiled or scattered food will. Food left on the ground is a sign you're over feeding them. Limit the amount of treats fed to what can be eaten in the session. Small pots of grit and oyster shell and wheat for waterfowl should be removed overnight as well. Now's your chance to get them topped up ready for the morning. If drinkers or feeders have been soiled, give them a clean as well.

Try to spend at least ten minutes a day with your birds. This can be treat time or just time spent handling or encouraging them closer to you so that they trust you. Holding a treat of dried mealworms or corn in your hands will encourage most birds to come to you over time. Don't rush your birds. Chickens tend to be the easiest poultry to hand tame, although with time even the most skitty of species or breed can usually overcome their shyness for a few grapes or other treats.

You will need to make sure that the run and house are kept clean for your birds. You should set an hour aside each week to clean the house out and put down fresh bedding and nest box lining for your birds. Some poultry keepers choose to operate a deep litter poultry house system whereby you cover the old litter with a fresh layer sometimes with a dry disinfectant between layers to help avoid smells.

Although in winter months this may be tempting, it is not ideal for backgarden keepers with neighbours, as it will increase bad smells and it also makes it more difficult to notice any fly, mite or other pest problems within the house. It is better to get rid of all the old litter. In winter months, using a dry disinfectant will stop the problem of the house still being damp from cleaning come bedtime for the birds. Damp conditions can lead to illness within the flock.

On a weekly basis, you should thoroughly check over any fixed runs for signs of predators. Areas where there is digging or holes could show that a fox has been visiting or a rat has been trying to burrow in to get to the food. Check your food store over as well. We used a wooden chest to store food in sacks until the rats gnawed up through the base. When cleaning, it's an ideal time to check for damage to the house. A leak in the roof may do no harm to your birds but will cause the wood to rot and a loose board can become Mr Fox's doorway in.

Try at least once a week to handle each bird. Check them over for any sign of problems such as lice or unusual lumps and bumps. This will enable you to know your birds better and will make sure they are healthy, with any potential illnesses promptly treated.

Every month you should also make sure, especially in fixed runs, that there is not a build-up of waste. If there is, take an hour or so to clean out as much as you can and to turn over chippings or other run base material.

Poultry, as with other animals, get worms. It is a good idea to keep an eye on droppings for any signs of worms and if there are to worm the flock. Many poultry keepers choose to preventatively treat on a three or six monthly basis as with cats and dogs. There are many wormers available, with herbal wormers for those working to organic standards, although it is often thought that these do not clear the bird of worms as well as conventional poultry wormers.

## Winter Gardens and Poultry

Come autumn and winter, most backgarden poultry keepers notice how damaged the ground can become where they keep their bird. Rain turns the grass into a mud bath and the cold nights then turn the mud bath into an ice rink, both of which end up with you flat on your back with inquisitive birds staring down at you wondering why you haven't fed them yet!

There are, of course, solutions to the problems and these depend on how much space you have to keep your birds in. If you have a large garden that can be sectioned off, we would highly recommend you set this up and rotate your birds on a fortnightly or monthly basis, giving the ground a chance to recover and to help avoid the muddy puddles poultry, especially ducks, are so fond of creating.

A lot of people, ourselves included, don't have the space to rotate their birds or have a fixed run that cannot be relocated. It is also the case that by winter the chickens and ducks in a small garden have pretty much destroyed what was the lawn and this becomes the worst spot for mud baths.

Of course, much of the problem is caused by the number of birds being kept in relation to the space. Three hens in the average small garden will not cause any real damage to the lawn but if, like us, you have half a dozen hens along with eight ducks in a relatively small garden, then you will have the problems.

If this is the case we would suggest that woodchip becomes your winter best friend in the garden. Put a 3-inch (7.5 cm) thick layer on the lawn and in front of your coops, or the same in a run if you have one, to stop the area turning to mud. It can be washed down and raked over a couple of times a week to keep it evenly spread and free of droppings. If you have a concrete run or fixed surface, keep this clean and on an almost daily basis brush or sluice out excess water from the area.

For fixed runs, you may wish to consider rubber chippings which last for years and can be washed over and raked through without the need for replacement. If you do choose rubber chippings, make sure your hygiene management is of a high standard to stop the area becoming over burdened with parasites from the droppings.

Whether in garden large or small or a run, poultry need shelter, especially in the winter. Make sure there are one or two spots that are sheltered from both wind and rain, perhaps using plastic sheeting or old fencing to section off a dry area for them.

Come springtime you could either rake up and compost the chippings and reseed the grass lawn (you would need to keep the birds off while it takes to seed or they will just gobble it up for you) or you could put down fresh chipping and choose to use this as a year-round garden cover for the birds.

In colder weather, you need to make a few basic checks of your coop or shed so that your poultry is kept in dry, draught-free conditions. The waterproofing needs to be checked and any areas that are dripping water into the housing need to be sealed quickly. Poultry don't mind the cold but they do mind draughts and the rain dripping into the house at night so these are a top winter priority for them.

Although they hold body temperature well with their feathers keeping them toastie warm in freezing weather, they do need shelter together as this allows them to share their body heat. Most poultry are creatures of habit so once you break the bad habits and get them going in at night they will continue doing so.

You can usually tell if there are problems in your coop or housing if usually well behaved birds suddenly refuse to go in at night. This could mean there is a cold draught, water leaking or more likely the dreaded red mite in there.

Most poultry will be fine through winter, but those who have moulted late or ex-battery hens, who are not yet fully feathered, need extra checks to make sure they are warm.

Some people have been known to knit jumpers for their chickens to keep them warm in the winter, but unless you have a secure run we would not advise this as it makes them even easier prey for predators as they cannot fly as well to escape.

We check our birds over once a day throughout the year and, especially in winter, look out for any big changes in body weight, blue tinge to the wattle and comb in chickens and look at their feet and legs as mud can lead to skin conditions which will need treating.

In winter, a few tips given to us have been: to put Vaseline on the ear tips, wattles and combs of hens to help keep them warm; put extra bedding in the housing to help them snuggle down in the warmth (make sure you clean up droppings daily as well); or, if really worried about warmth at night, put some old carpet with plastic over on top of the housing (making sure not to block ventilation) and this will keep them very warm.

Once the weather becomes freezing you will notice in the morning that your drinkers and ponds freeze over and if very cold they can refreeze during the day. The drinkers can be wrapped up in a layer of bubble wrap to help prevent freezing and adding slightly warmer water in the morning helps to keep it unfrozen for longer. In winter, we don't fill the drinkers to the top so that, when the water freezes and then defrosts, the expansion doesn't split the drinker making it unusable. Proper drinker covers can be purchased and even electric heaters to keep the water liquid throughout the winter.

Raising the drinker from the ground, either hanging it or supporting on wood or bricks, will help as the air warms faster than frozen ground in the dawn. We've heard of people using the small greenhouse paraffin heaters under a metal drinker in very cold weather but we think you'd need to be very careful that the water doesn't overheat and naked flame is always a fire risk.

Ducks and geese need water in winter and making sure it is not frozen over is a priority. Again, adding warmer water keeps it unfrozen for longer, but it may be necessary to break it up through the day.

If you're not at home in the day to keep an eye out, place a small ball in the pond or paddling pool and as this floats about it helps to prevent the water freezing over. Often though the ducks and geese are in and out so much during the day it remains fine and only freezes overnight.

Come winter, poultry often enjoy warm treats to help keep

their body temperature up. Remember that treats should only be given late afternoon, as they especially need the nutrients from their main feed in the winter. We make up a mash porridge in the morning for them by adding warm water and a small handful of porridge oats to a large portion of their pellets and give them this to start them off for the day as well as feeding ad lib pellets. In late afternoon we scatter a handful of corn for them to peck at as this helps keep them warm overnight.

## Summer Poultry

Whilst we tend to worry most about our poultry in the winter, the summer brings it own set of challenges. Chickens, being originally a bird that would live in the lower levels of the jungle, prefer warmer weather although they cope admirably with the cold under their feather duvets. But those feather duvets can work against them when we actually have a blazing sun.

If they're free-ranging around a garden, then they'll naturally head for the shady areas. You'll find them congregating under a hedge or bush. The cover also makes them feel safe from attack from the skies. We know there aren't any eagles up there but they don't!

If you haven't got bushes for them to go under, then provide shade. Even a patio parasol or beach umbrella will do. You must give them somewhere out of the direct sunlight.

One small warning, entering the run wearing flip flops or sandals, especially with painted toenails, is not a good idea, as Cara found out. The hens were fascinated by them and clustered around pecking her feet. She returned with socks on!

Make sure the house is well ventilated on those warm, muggy nights. In hot weather you may well need to clean the house out more frequently to avoid smells and flies accumulating. Lice and mites thrive in warm weather so you'll need to be vigilant and prepared with powders, etc., for them.

If possible, move the house to a shady spot or use an old sheet supported on some bamboo canes or wood to provide a sunshade for the house. Keeping the direct sun off the house will help a lot.

Never let them run out of water. An extra drinker will only cost a couple of pounds and can save their lives. The galvanized drinkers are better in winter as they are less likely to crack if the water freezes but the plastic drinkers tend to keep the water cooler in summer.

Waterfowl naturally keep themselves cool in water and whilst they technically only need a bucket of water to dip their head in, a paddling pool or a baby bath or even an old bath full of water will vastly improve the quality of their life. You do need to keep the water clean. We've had to refill the bath three and four times on a hot summer's day.

This can become a real chore if there is a hosepipe ban in force. However, some water companies appreciate the difference between watering the garden and livestock. If a ban is imposed in your area it is worth asking if it applies to you providing for your waterfowl. Talking of hosepipes, ducks seem to love playing in a shower. Using a fixed sprinkler to generate artificial rain really makes their day.

## Moulting

Moulting is simply the way chickens and other poultry replace their feathers. It can be particularly dramatic in chickens and worrying when you first see it as they can appear to be going bald.

When birds go into moult they will lose feathers, usually starting with one or two feathers and on occasion can lead to loss of 60–70 per cent of the body feathers at the same time. The feather loss is rarely symmetrical or contained in one area and will have a very patchy look by the time the moult is through. The new feathers will start to grow through quickly, and with heavy moults you may notice new feather coming through on the first areas of the moult while the bird continues to moult in other places.

The moult usually occurs at the end of the summer. However, as with anything poultry related, you can find they decide to moult any time of the year and only if it is in winter will you need to keep a closer eye on your bird.

Replacing all those feathers can drain the bird's resources. Egg production will fall off or may stop altogether during the moult. Making sure they are eating and drinking properly during this period is important. Adding extra protein can also help, as well as adding Poultry Spice or a Poultry Tonic to their water. Poultry Tonic is a liquid supplement that can be added to the water for poultry when they need added nutrients due to times of stress such as moulting.

We have found that a little tuna or tuna cat food goes down a treat with our hens and ducks during their moults; our only problem being the fact they then try to break the gate down at cat feeding time to get some more!

If the bird moults in the winter and it is a severe moult, you should make sure they have extra bedding and are staying warm. Ex-battery and barn hens often come out with little feathering, and hen jumpers can be purchased to keep them warm in colder months. These can also be used for a hen in a severe moult come winter time. But remember that jumpers restrict their flying and therefore ability to escape predators so such hens must be in a secure run.

**Keep Your Birds Entertained**
In the average garden, free-ranging birds will find their own entertainment. They'll be scratching around, digging up bugs and worms, hunting down those slugs and snails for you. They'll make themselves dust baths to play in and find the most improbable places to perch and rest in.

If your birds are in a run, then you will need to provide some entertainment for them. Bored birds fight so making sure they are happy and having fun will enable them to have a better life, fewer injuries and also increase egg production for you.

They should always have places with shade and shelter from both sun and heavy rain or wind. Chickens, turkeys and quail all enjoy dust bathing so a deep cat-litter tray, old washing-up bowl or something similar should be provided with a sand/soil mix for them to use. Keeping this covered to prevent it turning to mud in winter months is also a good idea.

Using dangling CDs, leafy green vegetables suspended from the run roof or on sticks gives them something to peck at other than the other hens. A small ball in the run gives the birds something to peck and play with.

Variations in height appeal to chickens. A straw bale or even an upturned box they can jump onto will suffice or a tree branch which they can jump onto and perch on.

Ducks will mainly enjoy playing in their water bowls and baths, but they also enjoy a straw bale in a sunny spot to make into a day bed.

## Catching and Holding Your Poultry

One of the first things you will need to learn is how to hold your poultry safely so that you can examine them. The best way to learn is to ask the breeder to show you how when you pick up your birds. However, sometimes in the excitement of getting your new birds things can be forgotten so the following is a basic guide on how to safely hold chickens, ducks, geese, turkeys and quail.

Chickens are easy to handle, although sometimes the catching of them can prove a little difficult for the more flighty birds in the flock. The best way to catch them if they run from you is to grab from underneath taking their feet in your hand. To hold onto the bird, keep the legs between your fingers and support the weight of the bird on the upturned palm of your hand. See Plate 5. Keep the head towards the centre of your body or, more accurately, the bird's rear to the outside. That way there's always a chance you won't need to change your jumper! If the bird is in a panic and tries to flap, use your free arm to gently hold the bird to your chest, thereby securing it and preventing injury.

Often the hen will just crouch submissively when approached and you can just pick her up with both hands. Do be calm and matter of fact when catching your birds. Nothing panics any animal more than a noisy, arm-flapping person. If you are in a state, then there must be something very bad about to happen!

Ducks and geese can be more difficult to catch mainly due to their larger body size and weight. Unlike chickens, they should not be grabbed by the feet or legs. To catch them, herd them into a corner where you can more easily get hold of the bird. Although it may seem that the wings will not take the weight, the best way to take hold of both ducks and geese is to take hold of the base of both wings and to lift the bird onto a table or other flat surface to examine them. If the bird has a tendency to bite or snap, you should use a restraining hand around the neck, but be careful not to over exert pressure. You can also lift ducks and geese by their body but again should be careful not to hold over tightly, especially with females, as it can cause damage to the bird. If you choose to body-hold the bird, keep both wings close to the body and make sure you support all of the bird's weight. See Plate 6.

Turkeys should be held in a similar way to ducks and geese. This is again due to their larger body weight. However, while holding the base of the wings you should also support the body weight with your other arm to avoid any injury to the bird.

Quails are best caught by taking a gentle hold of their body, keeping their wings to their chest. As they are small birds, be very careful about the amount of pressure used. If they begin to panic, it is best to release them into a dark box to allow them to settle, then to re-catch them once they are calmer to examine them.

**Wing Clipping**
It is a good idea to get your bird wing clipped by the breeder when you buy it, if necessary. However, you will need to re-clip after each moult. Again, ideally you should get the breeder or another experienced poultry keeper to show you how this is done, but in the event you cannot, it is in fact a fairly simple process.

When wing clipping a flighty bird, the purpose is to unbalance it, so you only wing clip one wing. We tend to always clip the left wing, thereby preventing any mistakes and accidentally clipping both sides.

You should only cut the primary flight feathers and should use sharp scissors and make sure you have good light when carrying out the job. Trimming these feathers does not hurt the bird; it is like having a haircut. However, you should make sure you do not cut too low in case you cut into a blood vessel. In order to avoid cutting too low, hold the wing to the light, note where the blood starts in the feather shaft and make sure you keep at least 1 inch (2.5 cm) from this point. Usually you will need to take only 3 inches (7 cm) off the wing to unbalance the bird in flight. See Fig. 2.

Once you've done it a couple of times, it's a job that takes only a few seconds. It's that first time when, despite what you have seen and read, you are scared of hurting them that is difficult.

**Fig. 2** How to clip a bird's wing – cut the first 8–10 primary flight feathers.

# 9

# DISEASES AND PROBLEMS

Thankfully many of the diseases and problems we list are unlikely to trouble the small-scale backgarden keeper. Purchasing healthy, clean birds and being careful when introducing new birds by observing a quarantine period means that there is little likelihood of them catching a disease.

You also need to be careful when visiting other poultry keepers or being visited by them. Some diseases are carried in the droppings and these can attach to shoes or clothing. It is worthwhile and only takes a little effort to clean the underside of your shoes and to wash your hands and so forth before going back to your birds after visiting other keepers or before visiting another's flock.

As mentioned in the chapter on Preparing for Your Poultry, we recommend that you find a suitably qualified and experienced vet at the outset as it can be difficult to find one at short notice, particularly in an emergency.

## General Lice and Skin Mites
Lice, depluming mites and, in the worst case, Northern Fowl mite can live in the feathers of the birds. While chickens, for example, dust bathe to remove these and ducks will water bathe to remove them, some may still remain. If you suspect lice or mites, part the feathers and look for little grey specks scuttling for cover or tiny eggs laid at the base of the feathers.

Lice and mites will, without treatment, lead to sick birds

that lose condition and slow down on egg production. There are powder treatments available from most poultry suppliers to help and often vets will prescribe a spot-on that is licensed for other animals to kill off any offending parasites. Preventatively dusting your flock on a monthly basis will help stave off any problem infestations. It is a good idea to make sure the house is thoroughly cleaned and dusted as well to kill any mites or lice in the bedding.

**Scaly Leg Mite**
This is a problem mainly associated with chickens, but other species can suffer and this is especially the case in mixed flocks. The mite causes the scales to become swollen and sore to the touch, often with a noticeable heat in the affected area. Special spray treatments can be bought that will kill off the mite. Some keepers choose to use an alcohol rub on the leg and Vaseline to suffocate and then calm and prevent repeat infestation during treatment. However, we found the specific treatments to be more effective and to cause less distress to our birds when we suffered an infestation.

**Red Mite**
A major bane for poultry keepers and one of the most hotly debated topics on most poultry keeping forums. In the past, many keepers swore by the use of Creosote in the poultry house to prevent infestations. However, due to legal issues, this is no longer an option for the backgarden keeper. Red mite are small mites, about half a millimetre long, that live in the poultry house and come out at night to feed off the birds' blood. They are grey in colour in the day when they haven't fed but red after they have fed due to being full of the chickens' blood.

It is often difficult to see them in the day since they are tiny and grey, and hide in crevices and cracks in wood. Whilst they cannot live on people, sometimes the first you'll know about them is an itchy feeling after going into the poultry house at night.

Red mite can kill birds if left for long periods, as the birds will become anaemic due to the infestation and, therefore, more prone to infections. One of the first signs of a red mite infestation is often the birds' reluctance to go to bed at night time. You can check by going out at night with a torch and piece of white paper. Run the paper around the house, especially perches and sleeping areas. If you have red mite, there will be blood on the paper indicating the mite.

Once red mite take hold, it can be hard to stop them, especially in summer months. You will need to clean the house thoroughly with something such as Jeyes fluid and air the house out. Then spray with Poultry Shield.

Dust with a red mite powder, or Diatomaceous Earth (Diatom) best painted on as a slurry, into all the nooks and crannies and use fresh bedding. Make sure you clean and puff dust into any nooks and crannies where the mite are likely to be hiding. It will take a number of weeks of treatment to eradicate and in the meantime keep a close eye on your flock to make sure they are eating, drinking and are not showing signs of listlessness or illness.

One of the benefits of the plastic poultry houses is that they give the mites fewer places to hide with smooth, impervious walls. Even so, a plastic house doesn't offer complete immunity from them.

**Worms**
As mentioned earlier, poultry will suffer from worms. Roundworms, hairworms and tapeworms live within the digestive tract of the birds and are most commonly noticed when excreted by the birds. Gapeworms live in the lungs of the birds and can cause breathing problems. They are often first noticed when the bird gasps for breath by repeatedly opening and closing its beak or bill whilst stretching out its neck.

Worms will cause the birds to be listless and will reduce egg production, as well as causing an increase in problems such as soft shells. The most common poultry wormer used is Flubenvet and, although not licensed for use with all poultry, is the wormer most commonly prescribed by vets for treatment. It is licensed for chickens and can be purchased without a prescription.

Always follow the advice on the wormer or that given by the vet treating your flock, especially in relation to dosage. Most wormers are added to the feed. With smaller flocks, it is often easier to dose enough feed for either five or seven days depending on the treatment time or to give each bird the wormer separately, for example, dosed onto half a grape for each bird.

## Coccidiosis

These are parasites that feed upon the birds from the inside and can lead to death. There are different types of coccidiosis and different poultry types are affected by different types of coccidia, so they are species-specific. It is more common in birds between 3–6 weeks of age. The symptoms are that birds become listless, pale and shiver as if they are very cold. Food consumption drops and often they have diarrhoea. If you suspect coccidiosis, you should seek a vet's advice and treatment for your flock as it can quickly kill, especially in chicks and younger birds.

A coccidiostat added to the feed for chicks helps to prevent coccidiosis, but medicated chick feeds should only be used for chickens and not for other poultry as the coccidiostat for chicks is harmful to ducklings, goslings and other poultry.

## Mycoplasma

This is a bacterial infection that affects chickens and other poultry. Its symptoms are sneezing, gasping, runny eyes and heavy or hard breathing, as well as a general loss of condition and listlessness. Mycoplasma requires antibiotic treatment and if your flock suffers these symptoms veterinary treatment should be sought.

## Blackhead

This is a condition that chickens may have without showing symptoms but pass on to turkeys in whom it is far more serious. See turkeys, page 132.

## Avian Flu – see Zoonoses (page 98)

## Aspergillosis

A disease usually found in a duckling that affects the lungs presenting with breathing difficulties and usually caused by mould being inhaled by the ducklings either at hatch or from mouldy bedding. It usually occurs within the first 5–10 days and if a duckling suffers from obvious breathing difficulties veterinary treatment should be sought as soon as possible. Those ducklings affected will need to be placed into a separate brooder to prevent infection of the rest of the hatch.

## Mudballs

Especially in winter wet weather, small particles of mud and bedding form into a ball on the underside of the chicken's foot which builds up into a larger and larger ball, making it difficult for the chicken to walk.

Don't try and just pull it off; you'll certainly tear some skin off if not do worse damage to the foot. You need to soften the mud with warm water and carefully pick it away, a bit at a time. It is one of those problems that the sooner it is noticed, the easier it is to resolve.

## Bumblefoot

This is a foot infection problem with chickens that can become serious and if left untreated cause the death of your bird. Bumblefoot often presents itself as a small red swelling on the base of the foot. It may also be shiny or, if the skin breaks with the swelling and irritation, the area may bleed or scab. The chicken will often limp or not rest on the infected foot. If left untreated, the area will become worse and puss may become present. This infection can cause distortions and disfigurement of the feet and toes, causing future difficulties with walking, perching and in cock birds with mating holds. Untreated it will probably result in septicaemia which is fatal.

The chicken's perches can cause Bumblefoot or the way the chicken is perching in its coop. Your chicken's perches need to be clean and smooth, free from any splinters or jagged edges that may damage the foot.

The perches should not be too high, because when the bird

jumps down it may injure its feet. Overweight or heavy birds should be monitored, as they are more likely to injure themselves when jumping from their perches. Another cause can be sharp objects within the run area piercing the chicken's feet. This allows bacteria to enter the foot, causing the Bumblefoot infection.

When building your coop and run, make sure there are no sharp or jagged edges or flooring that your birds' feet could be damaged on. The perches within the coop should be just above the height of the pop-hole opening – ideally around 16 inches (40 cm) off the ground.

The perch itself should be around 2–3 inches (5–7 cm) wide and 2–3 inches (5–7 cm) deep, with the edges rounded off and smoothed so that nothing catches on the chickens' feet while they perch. If you have bantams, reduce the perch size by half an inch (1 cm) as they have smaller feet and therefore smaller grip.

You can add higher perches if you wish, but make sure the bird can jump down easily by adding perhaps a lower perch to jump to, or a lower shelf before the floor height, so that injuries aren't caused.

As this is a bacterial infection, it can be treated by the use of antibiotics. It is recommended that the chicken be taken to the vet for the foot to be examined and a course of antibiotics to be administered. Treatment should be obtained quickly as this infection can soon cause further problems for the chicken. Extreme swelling sometimes leads to surgery being the only treatment option.

The foot should be disinfected by using an antibacterial animal spray or an iodine solution. Whilst the foot is healing, the chicken should be kept in a controlled area with clean soft bedding to avoid further irritation on the foot. The infection usually clears up within 7 days, but if concerned you should return to your vet.

There are home remedies that some people claim to help relieve the infection without needing to have vet treatment. These include bathing the area and disinfecting as described above, then applying an ointment of either Calendula cream or Comfrey cream to the area, morning and evening, to cleanse the infection and reduce the swelling, allowing it to naturally heal.

One experienced chicken keeper on our poultry forum suggested:

"Clean the foot completely with warm salty water and look for the puss spot. Lance it with a modeller's knife or scalpel and squeeze out the puss if you are able or just tease out the head with the scalpel point or cotton buds.

Put some salt on it for a minute or two and then rinse off, dry well and then apply Germolene antiseptic cream twice a day on a fabric plaster. Wrap the foot up as best you can with more plaster strip (only where it is affected by the Bumblefoot), cleaning the wound each time you redress it as before. Make sure the bedding is clean where the bird is kept. You may make the infection go but the swollen bit often stays for longer."

Although some chicken keepers swear that home methods of treatment are effective, if you suspect the infection is worsening or not improving, seek advice from a vet so that the chicken is not suffering.

**Marek's Disease**
This is a very serious viral infection that causes weight loss, paralysis and death. It is impossible to cure and vaccination is the only way to prevent it. Even vaccinated birds that are healthy in themselves can be infectious. Once in the flock, without vaccination, the death rate is typically above 80 per cent.

The rule is to be very careful where you obtain your poultry from and stick to reputable breeders. Vaccination is typically carried out on day-old chicks or by vaccinating the eggs with large-scale suppliers.

One problem for the small-scale breeder with vaccination is that the vaccines are only available in large quantities. When hatching half a dozen eggs, buying in a thousand doses of a vaccine hardly encourages the practice.

**Newcastle Disease**
This is viral disease that affects a number of bird species and can cause flu-like symptoms and conjunctivitis in people. In

poultry it is very serious and incurable. Typically the bird will have laboured breathing, followed by paralysis and death. Other symptoms may include: muscular tremors or shaking, twisting of the head into odd positions, swelling around the eyes, malformed eggs, reduced appetite and greenish, watery diarrhoea.

If you reasonably suspect you have the disease, then it is important that you contact a veterinary surgeon as soon as possible, even if your bird has died. He will assist you in reporting the disease as is required by law under the Animal Health Act 1981.

The law states, "Any person having in their possession or under their charge an animal affected or suspected of having one of these diseases must, with all practicable speed, notify that fact to a police constable."

In practice you (or your vet) will notify the local Animal Health Office. Contact details may be found online or in the phonebook but most vets can give you the number.

## Crop Problems

The chicken's crop is located right beneath the neck against the breast and just right of the centre. When a chicken eats, the food goes into the crop, which extends to accommodate it and, especially with young chicks, can be easily seen protruding after the bird has eaten. The crop is like a store where the food starts to soften before being passed down to the small stomach and gizzard where the food is broken down and digested. It is a vital organ and when keeping chickens you need to keep a close eye on your birds' crops to make sure they are working properly. A well working crop should be empty when the chicken wakes in the morning, there should be no strange smells emanating from the crop and the crop should fill during the day while the birds eats.

The two most common crop problems are sour crop and impacted crop. Impacted crop is also known as being crop bound.

Sour crop is caused when the crop doesn't empty fully overnight and as a result the food ferments within the crop causing a fungal infection. You can identify sour crop by checking the crop before the bird eats in the morning: if it is sour crop, the crop will be watery or squishy like a balloon and when you open the bird's beak a foul smell will emanate from it.

Impacted crop (crop bound) is similar in that the crop does not empty overnight but in the morning it will feel hard and swollen, like a golf ball. In both cases the bird is likely to seem lethargic, will lose weight and may make strange head movements due to the discomfort of the crop.

In order to avoid crop problems you should make sure that, along with pellets, there is an ample supply of poultry grit available for your chickens. The grit breaks up the food in the gizzard; without it, as the food cannot be broken down and digested, the food will build up in the crop.

Long grass should be cut down as this can compact in the crop and stodgy foods such as bread and pasta should be fed sparingly as treats as these can also cause a compaction. In addition, on a monthly basis, add Apple Cider Vinegar (ACV) to the water supply and make sure your birds are fully wormed as this also helps to keep the bird healthy and prevent health problems.

To treat sour crop, you should start by holding your chicken upside down with the head away from you and gently massage the crop from bottom to top so that the fluid is released. Only hold the bird like this for a short time, 15–20 seconds, to avoid choking her.

Once this has been done, mix some natural yogurt into the layers pellets or mash and feed this to the bird, along with water mixed with ACV throughout the day. You will probably need to repeat this over three days until the crop stops filling and the bacteria in the crop returns to normal. In the event the problem persists for more than a week, seek a vet's advice, as a course of anti-fungal medicine may be required.

If you suspect your chicken has an impacted crop start treatment in the morning by dropping a small amount of olive oil or liquid paraffin (2–3 ml) into the bird's mouth and then gently massaging the crop to help break up the compaction and repeat this procedure in the afternoon. Feed a soft food such as layers mash and again add some natural yogurt to soften the food and neutralize any bacterial build-up in the crop. It may take a couple of days to break up the compaction, but if it persists for three days the compaction may need surgically removing.

If you are not absolutely confident carrying this out yourself, take the bird straight to the vets where they can perform the operation.

In the event you wish to carry out this operation yourself, you will need a scalpel, rubbing alcohol, cotton wool, clean towels, water, saline solution (which you can buy at the chemist), an additional pair of hands to hold the bird securely and superglue.

Firstly wrap the chicken in a towel and hold securely. Next, locate the chicken's crop and sterilize with the alcohol and then make an incision of about 1 cm (½ inch) about half way down the crop, using enough pressure to cut through the skin and the thin muscle area below.

Once the incision is made, quickly remove the contents of the crop and then rinse the crop with the saline solution. Put a thin line of superglue along the incision and hold the edges of the muscle and skin together so they seal. Incidentally, super-glue was developed for the US Army for treating battlefield injuries and that is why it is so good at sticking skin together, including your skin to the chicken, so be careful.

Clean the exterior area and again rub with the alcohol solution. Isolate the chicken into a clean secure coop and do not give any food or water for at least 18 hours. After this, provide water and small amounts of liquefied food such as puréed fruit or mash. If you suspect infection, go straight to the vet for further treatment. Always seek the advice of a vet if the condition worsens.

**Feather Pecking and Cannibalism**
This can be a major problem, especially for flocks kept in runs. It mainly affects chickens and quail, not so much water-fowl. The best way to avoid feather pecking and bullying problems is to give the birds as much space as possible. Often a feather-pecking problem will start when new birds are added to the flock while the hierarchy is reasserted or due to

boredom within the flock. Another cause can be a vitamin deficiency in the flock. The problem is that once feather pecking begins, it can be difficult to stop.

If you notice that there is a pecking problem, you need to identify the "bully" and the main victims. Those birds that have been pecked should be checked over and the pecked areas should be covered with purple antiseptic equine spray. This will disguise the area as well as helping to prevent infection.

If blood is drawn, this can lead to a frenzy by the pecker (and other birds in the flock) and, especially with quail, can quickly lead to an entire stripping of the head or tail feathers (the most commonly attacked area) and can also lead to cannibalism and the death of the bird.

You should try to resolve any possible vitamin deficiencies by providing the birds with a diet of layers pellets, greens and oyster shell or another source of calcium. Make sure that the birds have entertainment as mentioned earlier so that they are not pecking each other purely out of boredom.

If you have a persistent offender among your chickens, you can buy beak bits that allow the bird to eat and drink but prevent feather pulling. You can use these to break the behaviour and then remove them or use them continuously after a problem has presented. However, continuous use causes the beak ring to wear a grove in the upper and lower beak so that the beak meets again, making the bit less effective

With quail, it is best to separate the offender within sight of the rest of the flock for 5 days and then reintroduce her to the flock, hopefully broken of the pecking habit.

## Wet Feather
This is a problem with ducks and geese that is most commonly caused by a lack of clean bathing water. Waterfowl need to keep their feathers well preened to prevent water logging when they swim. If they don't have enough water to clean themselves in or are unable to preen properly, the birds will start to look bedraggled and the feathers look clumpy and wet even when they have not been near water. The problem is that once birds have wet feathers, they become more reluctant to preen and bathe, exacerbating the problem.

Ensuring the birds have clean bathing water and drinking water, as well as a good diet rich in layers pellets, greens and wheat, will help to prevent any wet feather problems. Making available dry sheltered areas for wet weather will also help as it is more prevalent in wet winters. Keep areas clear from mud, and provide dry areas to sit and sleep on both in the garden/run and a clean dry poultry house.

If a bird gets wet feather, you should first of all encourage it to bathe and preen as well as checking for any parasites on the bird that may be making it reluctant to preen itself. If necessary, you can gently bathe the duck, but make sure it is fully dry by patting with a towel before it goes to bed. Adding more wheat to the birds' diet, in order to assist the production of oil in the preen gland, can also help resolve the problem.

**Spraddled Legs**
This is a problem in young hatchlings where their legs splay or spraddle apart. Young hatchlings should not be kept on slippery surfaces in order to help prevent this condition. If you notice legs spraddling, lightly tie the legs together while changing the flooring surface. After 4–6 days this should hopefully have been resolved. If the legs continue to have problems, you may wish to consider culling the chick, as it is likely to lead to health problems as the bird grows.

**Egg-Eating**
This is a fairly common problem with chickens, and is something that can usually be easily dealt with. Often the problem will start when one of the chickens lays a soft-shelled egg and eats the remains afterwards. If it is a case that the chickens are only eating when a softie is laid, you should treat the soft-shell issue and the egg-eating will stop.

In the event you have an egg-eater who breaks through a hard shelled egg, you need to make a few checks. Firstly, make sure that there is at least one nest box per three hens. Too many hens laying in the same nesting box can lead to stress situations and egg-eating. Secondly, make sure that all the hens are healthy, wormed, lice- and flea-free and that they are laying without difficulty.

*(Plate 1)* Ex-battery hens on their way to recovery.
(*Courtesy of Fresh Start For Hens; www.freshstartforhens.co.uk*)

*(Plate 2)* The female cream legbar chick (to the right of the male) has
noticeably darker colouring when first hatched.
(*Courtesy of Jenny Tonkin Nature images/Alamy*)

*(Plate 3, right)* Cockerel with wattle and comb.
*(Courtesy of D. Hurst/Alamy)*

*(Plate 4, below)* Chickens roosting at night.
*(Courtesy of Sami Sarkis/Alamy)*

(Plate 5, above) How to handle a chicken properly, and (Plate 6, below) how to handle a duck properly.
(Courtesy of Cara Harrison)

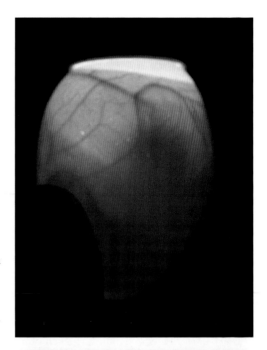

*(Plate 7)* Fertilized chicken egg. Candling reveals the blood vessels of a healthy 11-day-old fertilized egg. *(Courtesy of Science Source/Science Photo Library)*

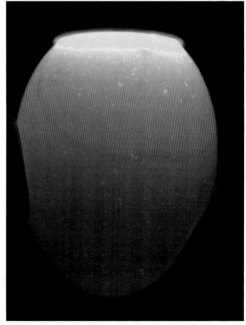

*(Plate 8)* Non-viable fertilized chicken egg. Candling reveals that the embryo has stopped growing, as shown by the characteristic thin blood ring (thin dark line) encircling the yolk. *(Courtesy of Science Source/Science Photo Library)*

If you have a healthy flock and plenty of nest boxes, you will need to make sure that you are removing any eggs at least twice a day so that they are not sat there for long periods, which encourages egg-eating. Placing ceramic eggs in the nest box can help to stop the egg-eater. Some people also recommend filling an egg with mustard and placing it in the nest box to deter any potential eating.

If the problem persists, you need to identify the egg-eater in the flock and separate her. Isolating her for two weeks and removing any eggs she lays immediately, as well as using the pot egg or mustard egg, should break her of her habit once and for all.

## Wounds

Minor wounds will generally heal of themselves without any treatment, but the problem, particularly with chickens and quail, is that the other birds seem fascinated and start pecking at the wound. So a minor problem can quickly become major. The purple spray is a lifesaver. Not only does it have antiseptic properties, preventing infection, but it colours the area purple, which disguises the wound so the other birds don't peck at it.

It is especially useful if you are rescuing hens as they will often be only partially feathered; it will help prevent any feather pecking when you first get your birds home until they feather up again.

When applying the spray, you need to be careful not to get any on your clothes as it will stain, and the same can be said for hands as we have on a few occasions wandered around with purple-stained hands, despite repeated scrubbing after treating a wound.

It is stocked in most poultry and equine stores, a small bottle will last you a long time and it proves a valuable addition to the poultry keeper's store cupboard.

If need be, isolate the bird for a few days in a run still within sight of the rest of the flock until healed. Obviously serious wounds or infected wounds may require antibiotics and a vet should be consulted.

Sometimes a wound will heal over but is infected and will swell due to puss accumulating. A traditional remedy is to mix some Vaseline with honey and apply to the wound. It draws the puss and acts as a barrier to further infection. Honey has traditionally been used as an antibacterial agent.

## Zoonoses

Happily there are few diseases that humans can catch from poultry and instances are rare enough to make the news. Please don't feel worried that you or your children are likely to catch something. You're not. However, we thought it appropriate to cover these. Basic hygiene, washing your hands after handling poultry with soap and hot water is the only precaution you need to take.

**Salmonella** can be a very serious illness, but is thankfully rare in domestic flocks. It does become, potentially, a much larger problem in high-density commercial flocks, hence their use of vaccinated birds. Hybrid hens bought as point of lay from larger commercial dealers and ex-battery hens will have been vaccinated against a number of diseases including salmonella. It can be present in eggs but the most likely method of transmission is via the droppings or from droppings on eggshell falling into the egg before cooking.

**Avian Flu** – At the time of writing there has been one scare about Avian Flu following a few cases in the Far East of human infection but it isn't generally something you need to consider. If there is a serious outbreak of Avian Flu in the West, then the main thing will be to keep your domestic birds under cover and isolated from wild birds and the droppings of wild birds.

The flu virus lives in many species and frequently mutates so circumventing the immune system of the host. Rarely the virus will jump species and so humans can be infected with a strain of flu that normally infects another species. If there is a serious outbreak then DEFRA and the government will issue advice and instructions.

**Newcastle Disease** is a serious and notifiable disease of poultry which can cause flu-like symptoms and conjunctivitis in people. Amazingly, there is research into using the virus as an anti-cancer agent because it can be tailored to attack cancer cells without damaging healthy cells. It's an ill wind that blows no good.

**Psittacosis**, which is normally associated with parrots and is also called parrot fever, can occasionally affect turkeys, waterfowl and chickens. The illness is spread in dried particles of the faeces and feather dust which is why it is associated with parrots that are kept indoors. It really isn't something that backgarden keepers need concern themselves with.

# 10

# BREEDING YOUR OWN

*We can see a thousand miracles around us every day. What is more supernatural than an egg yolk turning into a chicken?*
S. Parkes Cadman

One of the joys of poultry keeping comes from hatching and raising your own birds. It may be to increase your flock of laying hens, table birds or to try and make a small profit selling on young pure breeds.

Often you will find that many hens or ducks in your flock go broody at the same time, which can be problematic if you have limited space to raise the hatchlings, so good planning is important.

We know from past experience that the temptation to place eggs under a broody hen without having properly thought it out can cause problems and extra expense as you frantically search for a broody coop and equipment for the hatchlings due within a day or two.

Hatching can also be very addictive. Before you know it, you have three sets of eggs under hens and the incubators on standby waiting for all your new chicks and ducklings to arrive!

When it comes to hatching your own, the first consideration is what you intend to do with any male offspring. Keeping in mind that half of the hatch are likely to be boys, you need in advance to decide whether they will be (with sex linked

chicks) culled at hatch, raised for meat, kept as pets or sold/
re-homed.

The first two options are the most likely for small
backgarden keepers. Selling and re-homing unwanted males
can be very difficult. The main problem with cockerels is that
they are noisy and when they come of age their fights can be
vicious. Drakes, ganders and quail cocks need to be kept
separately unless within a large flock. If you have limited
space or close neighbours, you really have to think through
how you will cope if unable to re-home or sell them.

Culling a bird you have raised with the intention of
re-homing is very difficult. However, although not easy, it is
easier to cull a male bird you have had planned for the table
all along. The main thing to remember is never to name a bird
you plan to eat, although they should be given as much care
as the rest of the flock. Remembering on a daily basis their
final destination helps when the day comes for the evil deed
to be done.

So, if you decide you have the space and you know how
you are going to deal with the potential male offspring, you
should plan how you intend to hatch and rear the young.

**The Birds and the Bees**
First things first, however. When breeding your own, we need
to think about the birds and the bees. If you are breeding to
develop your own strain of poultry and have more than one
male, you will need to control the breeding or you'll not know
who the father is. Keeping the males separate from the general
flock and using a breeding pen is therefore required.

If you only have one male, then knowing the father isn't a
problem but some people have a dislike of eating fertile eggs.
We can't see any difference between a fertile and non-fertile
egg until they have started being incubated but it is something
you may want to consider.

Only breed from your best specimens and unrelated birds to
prevent reinforcing bad genes which will cause congenital
defects.

With chickens it is best to use a fully grown cockerel. An
adolescent bird may be dominated by the females, develop an

**Fig. 3.** The life cycle of the chicken.

inferiority complex and become impotent. We kid you not! For this reason, it's a good idea to bring the ladies to the cockerel rather than the other way round. He's confident in his own territory and they're less confident as they're in his patch.

With cockerels mating, spur trimming is a good idea or he may accidentally rip the sides of the hens. Some cockerels can be quite dominant and bad tempered, even attacking their owner, so spur trimming is sensible anyway. Rub baby oil into the spurs for three days to soften them and then trim the tip. Claw clippers, available from most vets for cats and dogs, are useful here. If the spurs are still too long, repeat the process but only trim a small amount each time or you may hurt your chap and he'll remember. You may hear that spur trimming reduces the male's virility. This is total bunkum.

Another option is to buy a poultry saddle to protect your hens from the cockerel's spurs.

Once mating has occurred, it takes about three days before the eggs will start being fertile. The hen stores the sperm in a special organ called the infundibulum where the developing eggs are fertilized. The sperm can remain viable for three weeks. If breeding from a hen which has been around another cockerel, you need to keep this in mind.

With waterfowl, whilst they can mate on the ground, they generally are more successful when on enough water for them to swim in, 6–8 inches (15–20 cm) deep. However, this can cause a problem with an overly amorous large male and a small female. The male holds the female by the neck and can drown her. It doesn't happen often, but is something to be aware of.

### Broody Hens

Using a broody hen is by far the simplest method. She will sit on the eggs to keep them at just the right temperature, turn the eggs regularly and then help hatch and raise the offspring.

Ducks and geese are less prone to broodiness although some duck breeds such as the Muscovy and the Appleyard are more likely to go broody and complete a hatch than others. Luckily the broody hen will look after any eggs placed under

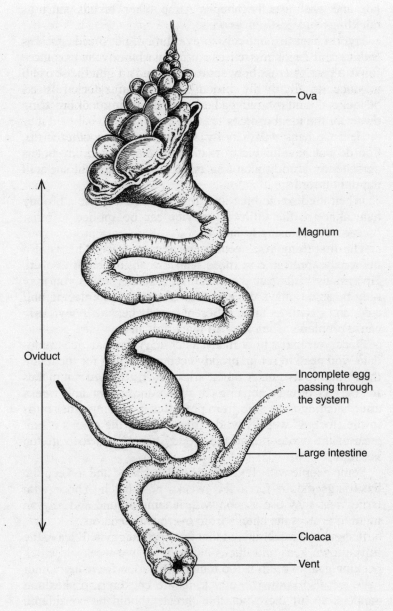

**Fig. 4.** The chicken's reproductive system.

her and will usually happily raise other breeds such as ducklings and goslings as well.

Hybrid hens are unlikely to ever brood, but breeds such as Silkies and Light Sussex can often go broody two or three times a year. A cross between a Silkie and a Light Sussex will produce an incredibly broody hen, often referred to as "Clockers", and if buying a hen with the intention of brooding these are the ideal choices.

Because hens will happily accept eggs from other birds, you do not have to worry that a hen that isn't in lay but is persistently broody won't sit on eggs. She will and she will thank you for it!

When it comes to breeding, most people will use a broody hen, although the following advice can be applied to other breeds you are using to hatch.

The first thing you need to consider is space. This is also obviously applicable to those who use an incubator as well. How many birds can you raise? In a small garden you may only be able to have three or four extra birds, so placing half a dozen eggs under three different broody hens will obviously cause problems when they all reach full size.

Once you know how many hatchlings you can accommodate, you need to set up broody accommodation for them and their mum to live in. A rabbit ark with an attached run makes for perfect broody housing. It allows those first few weeks after hatching to be safe from predators, including other birds in the flock as well as wild birds. It gives the chicks a safe taste of the outdoors and the broody hen a chance to stretch her legs.

Some people will choose to use a large shed and to keep the hatchlings indoor for a few weeks post hatch. Those with large areas may use a coop with fencing around and rely on mum to protect the chicks from overhead predators.

It should be remembered that your pet moggy will leave the fully-grown hens and ducks alone but a two-week-old chick pecking near to mum is too much temptation for many. Other hens are also known to attack, as are cockerels, drakes and ganders, so all these possible threats should be considered when you set up the broody house and run.

When it comes to selecting the eggs to go under the broody, this is all down to personal choice. The hen will sit on the eggs and a broody chicken will sit for 21 days for chickens, 28 for ducks or the 35 for Muscovy ducks and 30 for geese without any difficulty.

If setting six eggs it is likely only four or five of these will hatch, but never set more eggs than you can accommodate as that will be the 100 per cent hatch rate rarity! Remember that eggs must be fertile to hatch and, therefore, chicken eggs from chickens running without a cockerel will not incubate and hatch.

If you do not have a cockerel (or drake or gander), hatching eggs can be bought. The best place to buy is from a local breeder as eggs sent through the post can easily be damaged thereby reducing fertility. If you cannot buy from a local breeder, some known reputable breeders are happy to wrap and post to you; this is the second best way of obtaining the hatching eggs.

The third, and most risky, way is to buy from auction sites such as eBay. Although a lot of sellers are, of course, honourable there have been many cases of people being sold dud eggs or the wrong breeds through these sites, so caution should be used.

Eggs should be clean, free of hairline cracks and no older than 10 days old when being chosen for sitting.

Once you have got your hatching eggs you should let them rest for 24 hours at room temperature. During this time you should prepare the broody coop and area that is going to be used.

At night you can safely move the broody hen to her hatching coop and move the eggs underneath her. If she is sat on eggs already and these are fertile and you wish these to be hatched, do not add new eggs as well as they will have different hatch dates and this will lead to the mother hen leaving the nest before the hatch is fully finished in order to feed and look after the first hatchlings.

Once settled on the eggs, leave her until the following morning and check that she is still sat on the eggs and none has been kicked or removed from the nest.

If the hen is sat happily on the eggs in the morning, hopefully she has accepted them and will sit until they hatch. A broody hen will usually only leave the nest once or twice a day to defecate, eat and drink. Some broodies refuse throughout the hatch to leave at all, but you must make sure they do take a rest break.

Allow the hen fifteen minutes to look after herself twice a day and then encourage her back onto the eggs. If the area around the eggs or any egg gets soiled, it should be cleaned. If an egg is soiled and cannot be gently wiped off, the egg should be removed from the hen.

Sometimes the hen may kick an egg from the nest. If this happens and the egg is still warm, try to place the egg back into the nest underneath her. Marking each egg will help you identify them.

If the egg is kicked from the nest a second time, it is most likely infertile. Mother hens are very good at knowing if the egg is developing properly and can be trusted to remove eggs like this to protect the rest of the hatch.

After a week you may wish to candle the eggs to check fertility for yourself. This is a way to see what is happening inside the egg using a bright light. It used to be undertaken in a dark room using a candle with a reflecting curved mirror behind to concentrate the light, hence the name candling.

Purpose built candlers can be bought or you can use a small torch to do this. Very cheap Cree LED torches which are incredibly bright are available on eBay. You only need a small one. Ours has a facility to focus the beam tightly as well, which is useful.

You should candle the eggs at night so that you can see clearly and also so that the mother hen is not disturbed. At a week after starting to develop, a fertile embryo should appear as a small blob with blood vessels that radiate out from it. See Plates 7 and 8. If you cannot see the embryo, do not worry too much. Some coloured eggs are difficult to candle. Until you learn what to look for it can be hard to tell, so if the mother hen is still sitting we would recommend leaving the eggs with her. It may result in a failed hatch but that is better than discarding viable embryos at this stage.

10 days

20 days

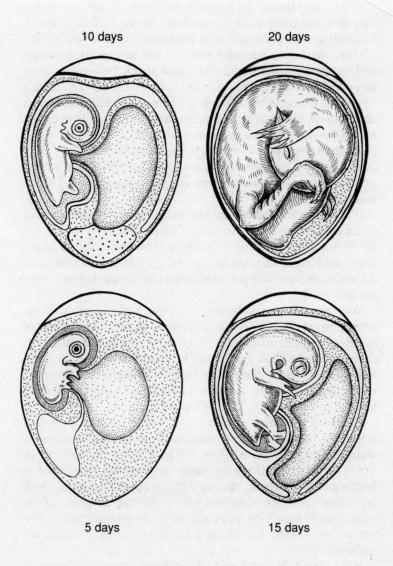

5 days

15 days

**Fig. 5.** Foetal development.

A few days before the eggs are due to hatch, the mother hen may be even more reluctant to leave the nest. This is due to her feeling the movement and impending hatch of the chicks.

Once the hen is adamant she will not be moved, leave her with food and water in the house and check a couple of times a day for hatch movement. Do not keep lifting the hen. Instead listen closely: you can often hear the pip-pipping in the eggs, chipping of the eggs or the peeps of the newborns snuggled happily underneath mum. If you hear nothing, lift the mum no more than once every six hours to see if there are any difficulties in the hatching. Take care when lifting the hen as they often tuck eggs up under their wings. These could fall and break.

It is advised by many breeders that a chick that cannot hatch on its own or with just the help of mum should be left, as it is likely not to survive even if helped. It is obviously up to you if you wish to do this; we have to admit that on more than one occasion we have helped a chick that has struggled out of the shell.

With those we have helped, we have had a roughly 75 per cent survival rate. However, this is just our experience and all hatches progress differently. If you do choose to do this, you should peel the shell gently and be aware that there are many blood vessels so if there is any sign of blood place the chick in the egg back under mum and check back after 30 minutes to an hour to see how the chick is progressing.

Sometimes a chick may be born with a deformity. If with a mother hen, the mother may kill the chick. However, it may be that, for the welfare of the chick, you have to do this. Chicks suffer from the same illnesses as adults, such as pecking, bumble foot, etc, as well as spraddled legs. These are covered in the Diseases and Problems chapter. In the event you cannot cure the chick you must either take it to a vet or break the hatchling's neck so as to prevent any suffering.

After about 36–48 hours, the mother hen may choose to leave the nest with the hatchlings. Sometimes she will leave behind eggs and this can be because she no longer feels movement or she is concerned about the hatchlings that have already arrived now needing to eat and drink. If you suspect the egg is fertile or can hear or feel movement and have an incubator, you can try to finish the hatch yourself, although success on this can be low.

Some people choose to check the eggs left behind to see what went wrong. However, for novice or hobby breeders, we would say that disposing of the eggs and concentrating on those new birds with mum is the best option.

Mother hen will take the chicks to the chick crumb and water and teach them how to peck and how to drink. If you have had a mother hen hatch waterfowl you will have to wait a little while before giving them swimming water. When waterfowl hatch, their mothers preen them giving them their waterproofing, but with incubation or surrogate mother hens they do not get this, so they need time to develop the waterproofing on their own before entering the water.

After a week you can put a shallow water bowl for swimming in with them. You will find that mother hens may panic at their chicks' desperate need to climb in the water, but after a day or two they get used to this strange behaviour by their offspring.

Although covered runs are best with young hatchlings, mother hens will protect their wards and after three or four weeks the chicks will be getting big enough to meet the rest of the flock. If the broody coop and run were kept in sight of the rest of the flock, this will be easier as they will already be aware of each other. Otherwise introductions should be handled carefully and you should be ready to intervene if the hatchlings are in danger.

After six weeks, mum should be given more layers pellets and the hatchlings moved to a growers pellet. They will be getting quite big now and mum may seem less interested. She may also try to take them back to the main coop with the main flock.

This is obviously fine if the flock has now integrated but if a mother hen has hatched ducklings or goslings they will not be able to perch and you may need to either encourage mum to stay in the broody coop or, if the weather is warm, keep them in the broody coop on their own at night.

We have found that when hatching ducks with a mother hen, the mother leaves them at six weeks. However, by this stage the youngsters have usually tried to take themselves to other ducks if kept in a mixed flock and that, by 8–10 weeks, they can be moved in with the adult duck flock.

## Incubator Hatching

Using an incubator is obviously more complicated than leaving Mother Nature to carry out the hatch for you but there are some advantages to having an incubator. You may not have a broody hen for starters. Some broodies are better than others; if your broody decides to stop incubating, then you have a back up with the machine.

When it comes to selecting an incubator, it can be very confusing with the large range of machines on the market. For home hatching you would be best settling with something that will set up to 24 eggs but you may feel 10 eggs at a time is more than enough. Humidity control and accurate temperature control are a must and automatic turning a real boon.

As with any equipment, it is worth going for quality and buying the best you can afford. You may not think that you need a large incubator for 24 eggs but this is an addictive hobby and in a couple of years you may well find yourself regretting buying a smaller model. Often they come with different-size egg dividers, which are useful if you wish to hatch, for example, chickens and then quails afterwards. Most incubators come with a hatching guide which you should follow.

The eggs should be fairly fresh; after a week the viability starts to fall. They also should be clean. Before putting them into the incubator, wash them in lukewarm water to which an egg sanitant has been added. The incubator itself should be scrupulously clean and sterilized.

## Temperature

The optimum temperature for hatching hens' eggs is 37.5°C for the first 18 days and then dropping by half a degree to 37°C for the last few days until hatching.

## Humidity

When the egg is laid, there is a small air bubble under the rounded end separated from the contents of the egg by a membrane. This serves to relieve stress in the egg caused by temperature changes making the contents expand and contract. This air bubble grows with time.

If the air is really dry, the fluid in the egg evaporates more quickly than it should, expanding the air bubble but not leaving enough fluid for the developing chick. Even if it does develop correctly, it may become stuck to the dry shell when hatching time arrives.

Conversely, too much humidity and the air bubble will not expand enough so that, come hatching time, the chick attempts to break through the shell in fluid and may drown. When breaking out of the shell, the chick breathes the air in the air sac within the shell.

Ideally hens' eggs should have a humidity of 52–55 per cent for the first 18 days and be increased to around 70–75 per cent thereafter until hatching.

## Turning

Turning the egg is especially critical in the early stages. If the egg isn't turned, the embryo will stick to the membrane inside the shell and become deformed. In the early days the egg should be turned four times a day at least, preferably six or more but turning should be stopped for the last four days before hatching. It's very easy to forget to turn the eggs resulting in a failure.

If manually turning the eggs, then mark them so you can follow the rotation. The egg should be positioned on its side with the rounded end slightly higher as if lying in the nest.

## Process

You will need to monitor the eggs much more closely than with a broody hatching them, especially watching the temperature and humidity as above, as well as the condition of the eggs.

Candle the eggs after a week and re-check any you're not sure of after 10 days. These should be removed before they go off. Don't forget hygiene: wash your hands with antibacterial handwash before handling the eggs. Any eggs that swell, sweat or smell should be removed from the incubator immediately as these may explode and spoil the whole hatch.

When you are three days from hatch you should change your incubator's setting to "hatch" or move to a separate hatcher. Do not turn the eggs after this point.

Once the eggs have hatched and the chicks have dried, becoming fluffy, you should move them to their brooder. This can either be a purpose-built brooder or a home-built brooder. We have in the past, in an emergency, used a cardboard box with lamp and a 100 watt bulb as a makeshift brooder. A red light (infra-red) is less stressful and you may have problems even finding a traditional 100 watt bulb nowadays as they have been replaced with energy-efficient bulbs.

Ideally, you will have a proper heat lamp. You can tell if the hatchlings are happy as they will be spread out happily around the brooder, pecking and playing. If too hot, they will be around the edges; if too cold, they will huddle under the heat source.

The hatchlings will need to be kept in the brooder for at least four weeks. Reduce the temperature gradually by about 2°C each week by raising the lamp. Sudden temperature changes will kill young chicks. In nature, the mother hen takes care of this but with incubation it is up to you to make sure that the chicks are warm as well as fed and watered.

You should find that the hatchlings take to eating and drinking with no problem. Make sure they have access to water and chick crumb at all times. Never let it run out. Shallow bowls are best; never have a deep bowl with water in or you may find a drowned chick. Chicks seem to have an inexplicable urge to get into water, especially when very young. To prevent accidental drowning, check that the water level isn't deep enough for youngsters to submerge their heads by popping some glass marbles or clean pebbles into the water. The marbles or pebbles can be washed easily. Chicks also poo a lot, so frequent cleaning of the brooder and bowls is a necessity.

From personal choice we would always opt to use a mother

hen over an incubator for ease, but you will find that hand-raised chicks are more attached and often more friendly than those raised naturally. For example, Cara has two ducks that she hand-raised that follow her around the garden and that actively want to be picked up and sit on her knee for hand feeding.

If keeping a cockerel, they are also often better hand-raised. Cara has had one Poland cockerel that used to ride around on her shoulder and was softer and more friendly than any hen she has had since.

## Sexing Chicks

Once your chicks hatch, one of the first things you will want to know, after checking they are all healthy and happy of course, is what sex they are. The likelihood is you will have a pretty much 50:50 male to female hatch. However, sometimes you may be lucky and get that elusive 100 per cent female clutch; and, at other times, a 100 per cent set of males.

The easiest way to know on hatch what you have got is only to incubate or set under your broody autosexing breeds. These include breeds such as the Cream and Gold Legbars who can be sexed on day one due to the coloration of their feathering (the Cream Legbar girls being darker with a more defined head spot than the boys).

Another method of sexing with some breeds is to note the coloration of their wing feathers as they develop at about 10 days old. For example, a Salmon Faverolle female and male will be very different, with the male having black feathering developing and the female a lighter salmon coloured wing.

If you hatch a breed that is not autosexing or does not have identifiable plumage, you will need to wait a few weeks to sex your birds unless you learn how to vent sex. Vent sexing is very difficult and should not be tried by a novice keeper as it can cause internal damage to the bird. As a backgarden keeper, as hard as it is, it is best to wait for other signs of sex development.

At about 2 days old you can attempt to feather sex some breeds. The chicks will develop their first wing feathers at different rates, with the male chicks developing their primary and secondary feathers at a similar rate whereas the females will develop their primary feathering faster than their secondary, giving the effect of a faster developing wing. This only works until about 4 days old and some breeds such as Silkies do not develop in this way.

With your male offspring, you will find that their wattles and combs develop faster than the hens in the hatch. At about 3 weeks, the cockerels will have a more defined comb that will already be starting to colour up with most breeds whereas the hen will still have a smaller paler plume.

By about 8 weeks of age, you should be able to distinguish between the boys and girls in the hatch by their size, wattle and comb, tail length and to some extent by their behaviour. We have found that at least one male will start to become the boss even at an early stage (a week or so after hatch).

Of course, the most definitive way of knowing which are boys and which are girls is to wait until the day they crow or lay at about 18–22 weeks of age.

# 11

# DUCKS

Ducks in the garden are becoming a far more common sight in urban neighbourhoods as they regain popularity with the backgarden keeper. They are a real joy. They can be easily kept to an area without the need for high fencing, as backgarden breeds are largely flightless.

Although a little messier than chickens, their actual care needs are very similar, and with food, water, good housing and some fuss and attention, they become loving additions to the backgarden flock.

## Ducks to Water

Ducks, of course, have one major need that differs from hens: bathing water. A lot of people just don't consider ducks as they think they need a large pond which few of us have room for.

The answer is to pick suitable breeds. Backgarden keepers find that a small paddling pool or even large buckets of water will be enough to keep their ducks happy and healthy. But beware: the area around the water will quickly become muddy so placing pools on well draining ground is a must.

We use the large, hard, plastic paddling pools as well as a large, deep bucket and a bucket drinker. In the winter we put one of these pools out and in summer months two pools as they need extra water to splash in. We change the water on a daily basis.

Some of the larger breeds or older ducks may need a step or ramp into the pool. Young healthy birds have little trouble hopping in and out of the pools as well as happily splashing anyone who is silly enough to stand too close!

It is important to ensure that ducks have at least enough water to enable them to dunk their heads in. This helps them with the preening process and to keep their beaks and nostrils clear and clean.

It's a good idea to plan where you will drain the water. In Cara's first year of duck keeping she made the mistake of standing the ponds on soil in the garden. In the summer months this wasn't a problem but in the first winter a muddy little swamp developed. The ducks seemed to love the swamp, but it wasn't the best for their health and feathers. The ponds are now positioned on a concrete hard standing with a slope off for drainage. This means that on emptying the ponds in an evening, the water can be swept off the area to drain. This keeps the area clean and dry even in the winter months.

## Duck Eggs

Like hens, some breeds of ducks will lay only a few eggs while others are excellent egg-layers, laying as many as any breed of chicken. In contrast to hens' eggs, duck eggs are generally larger and have a larger proportion of yolk to white.

Using duck eggs when cooking can be great for the baker in the family as they are reputed to make better cakes and are particularly good for scrambled eggs and omelettes. There is a slight difference in taste and consistency between a hen's egg and a duck egg. The duck egg has a slightly stronger, more eggy flavour. Because the duck egg contains less water, the white is noticeably firmer and can be a little leathery when fried.

Often though, those who dislike (or say they dislike) duck eggs do so through lack of trying. Even nowadays with chefs using different species eggs and with the wider availability in shops, many people are reluctant to venture away from the conventional hen's egg.

From March through September, Cara's six girls lay 4 eggs a day (the girls alternating their off days). This peters off over the winter months. If using no additional light for your flock, you will find that egg-laying slows through October and November, and that December and January are often barren months, with egg-laying slowly coming back in February.

**Ducks and Drakes**
Unlike chickens, the males (drakes) make less noise than their female counterparts. In the backgarden setting you can keep a drake without upsetting the neighbours. Ducks often outlive chickens by many years (some eight or so years) and for the final few years will not be productive, which is something that should be considered before venturing into duck keeping.

As with hens and cockerels, you do not need a drake to get the ducks to lay eggs unless you want them to be fertile for breeding. One drake needs at least three females around to keep him happy, the sexy beast. If you have more than one drake, you may have problems with the males fighting over the girls if you don't keep a high enough number of females per male though.

**Housing and Equipment**
The shape of the duck's bill is very different from that of the chicken's beak, so they prefer to shovel their food in rather than peck at it like a chicken. It doesn't stop them from picking up grain from the ground or devouring any slugs they come across. Our hens prefer the snails and the ducks prefer the slugs so between them they make a highly efficient, garden pest-control team.

Sturdy feeders and drinkers should also be provided as ducks seem to favour eating from bowls or troughs as opposed to traditional chicken feeders. Create a small covered area in the run for the feed to be put out in. The drinkers should be refreshed daily and the food taken up at night to avoid any rodent problems.

Ducks, unlike chickens, do not require perches and

basically need a large flat area that can be layered with
bedding, such as wood shavings or Hemcore, and straw-lined
nesting areas.

Most backgarden keepers choose to use a small garden shed
for their ducks as this gives them plenty of space and allows
for ground-level nesting areas. Cara has a large doghouse with
split doors. It gives them plenty of space and is high enough
to get in to clean with a spade. There are many options to
choose from, as long as you remember to keep the house at
floor height or to provide a ramp if it is raised off the ground.

One tip if you have a problem with rats gnawing holes into
the shed is to fix close meshed weldmesh to the base and first
foot (30 cm) of the shed wall.

In terms of space, small ducks (e.g. Calls) will need
about 1 ft$^2$ (0.1 m$^2$) of sleeping space, medium breeds (e.g.
Runners) 2 ft$^2$ (0.2 m$^2$) and large breeds (e.g. Aylesburys)
2½ ft$^2$ (0.23 m$^2$). If you are creating a pop-hole for the ducks,
make sure it is large enough for the larger breeds to get
through easily. It's actually better to open the door for them
than having a pop-hole, especially at night time when they are
herded back to bed, as they all try to run in at the same time!

Ventilation should, as with chicken housing, be above head
height. They will create a large amount of body heat at night
so there is no need to worry about how warm the house is for
fully grown ducks. If it gets cold, they will cuddle together; in
the warmer summer months they will spread out around the
duck house. Younger birds (under 10 weeks) if in with the
adults will be fine in colder weather, but if on their own you
may want to consider using a heat lamp in the colder months
to keep them warm.

You can try providing nest boxes for laying ducks but
they're notorious for laying where they will rather than where
you wish. We don't bother.

With enclosed runs, a 4 ft (120 cm) high fence will be
adequate to keep flightless breeds out of the vegetable plot.
However, the same rules apply as to other poultry when
considering foxes and other predators. You will need a tough
6 ft (180 cm) high fence with either an outward sloping piece
at the top or a roof above the run. Or use electric fencing, of

course. Even the large breeds like Aylesbury are vulnerable to the fox. Please don't take chances.

Giving them places to shelter is important, especially shady areas to keep cool in during the summer months. Although they do love wet weather, at times they will want shelter from the rain or snow so you must provide covered areas for them.

## Obtaining Ducks and Breeding

As with chickens, you have a number of options: breeding your own, buying hatching eggs, buying day-old chicks or buying young birds at point of lay. Don't forget that point of lay usually describes a bird aged from 16–20 weeks, and they won't actually start laying until 22–30 weeks, depending on the breed.

If you decide to breed your own ducklings, there are a few things to remember. Although ducks do go broody, they are not as reliable for sitting as many pure breed hens. The better broody ducks include the Appleyard and the Muscovy.

If you have a broody hen like a Silkie but want to hatch ducklings, you can place the eggs under the hen with no worries. Even though the incubation period is longer (28 days or 35 for Muscovy eggs instead of 21) the hen can feel the movement within the eggs and will happily sit for the full 28 days.

You can, of course, use an incubator as with chicken eggs but you will need to amend the settings to suit ducks. Most domestic utility breeds will need a temperature of 37.5°C and humidity of 58 per cent for the first 25 days and a humidity of 75 per cent at 37°C for the last three.

Muscovy ducks will need a different setting: 37.5°C at a humidity of 60 per cent for the first 31 days and 75 per cent humidity at 37°C for the last three days. Some ornamental breeds will need different settings again and the breeder should be able to advise you.

If brooding yourself you will need, as for chickens, a box and heat lamp. Do make sure the floor is non-slip as slipping on flooring can lead to problems with splayed legs. Treat as turkey poults with regard to flooring.

A broody hen will treat the ducklings in the same way as

she treats any chicks, once hatched; the mother's hen main concern tends to be when her new hatchlings attempt to jump into the water bowl!

A shallow dish for the chick crumb and another with water is best for the first week. If hatched by a duck, the ducklings will quickly gain the necessary waterproofing of their feathers from the mother preening them. However, if raised by a hen do not give them deep water to swim in for the first three weeks; although after the first week shallow water can be given so they can paddle in it.

## Feeding

For chicks from one day old when the yolk sac is exhausted to 3–4 weeks old, provide chick crumb specifically formulated for ducks and geese ad lib. Under no circumstances use medicated chicken chick crumb as the coccidiostat it contains is harmful to waterfowl.

From about 3 weeks old, start to mix in grower/finisher pellets with the chick crumbs to ensure a smooth changeover. They can be fed ad lib to growing birds until about 16 weeks, and for those being reared for meat they may be fed right up until finishing.

Once over 16 weeks of age, duck layers pellets can be fed ad lib, as well as cracked corn in the early evening. If you keep both ducks and a drake, the drake will be happy feeding on the layers pellets; and these provide the ducks with all the nutrients they need to lay healthy eggs.

If keeping ducks alongside chickens, you can allow all adult chickens and ducks to feed from chicken layers pellets with cracked corn in the evening, as the nutrient content of duck and chicken layers pellets is very similar.

Grit and oyster shell should be provided for ducks just as for chickens.

## Breeds

**Aylesbury** – the typical Jemima Puddleduck duck. They are large, white and with good handling can be very friendly, as well as being flightless due to their size. They make a great dual-purpose bird and the drakes can be dressed out to about

10 lb (4.5 kg). A lot of Aylesburys are no longer pure in the breed due to outcrossing with the Pekin and other varieties so when purchasing make sure you look for a pink/flesh coloured bill if you are interested in showing. The ducks should lay up to 100 large white eggs a year and also tend to go broody in the spring after their second year.

**Cayuga** – a large duck with black feathers that shine beetle-green in the sun. As the ducks age, they will develop white feathering but the drakes never do this. They are prized for their eggs, which in a good strain will be black at the beginning of the season. They should lay between 60–110 eggs a year in the spring and summer months. With regular attention they can be friendly and due to their weight are flightless once they reach maturity.

**Indian Runner** – these are a very easily recognized duck due to their upward carriage and long necks. They are medium weight and a very popular backgarden breed due to their flightlessness and their egg production. The ducks lay up to 200 large blue eggs a year. They come in many colours including blue, khaki and chocolate and are a joy to watch in the garden. Although they are suited to backgarden keeping, they are easily startled so not the best breed to choose with children or dogs.

**Appleyard** – a beautiful large duck with silver mottled feathering, lovely brown eyes and a beetle-green head in the males. They also come in a miniature version perfect for smaller gardens. They are friendly and inquisitive ducks and they also make good broody mums come springtime. The drakes make excellent table birds and should dress at 8 lb (3.6 kg). The duck will lay up to 180 large white eggs a year.

**Call Ducks** – very cute and attractive-to-look-at ducks, but they can be very noisy so not best suited to close urban areas. They are also a little more flighty than their heavyweight friends so wing clipping is a must unless they are allowed full free-range. There are many colours including white, pied and apricot. They will lay one, maybe two, clutches of eggs a year so about 12–20 in number in the spring.

**Muscovy** – a large duck with unusual bright red cresting around their bills and head. Our domestic ducks are all descended from the wild Mallard except for the Muscovy, which originated in South America as a distinct species.

The drakes can be quite aggressive and they hiss instead of the quiet male quack, especially in the breeding season, but the females are friendly and quiet so a plus if you have close neighbours to consider.

They come in lavender, blue, black, white and chocolate colouring and make great broodies. In fact, they can go broody repeatedly through the spring and summer which should be considered if you want egg production and have limited space.

Their eggs take longer than other duck eggs to hatch – up to 35 days – and their production is low. They are capable of flight so clipping or fencing should be considered if you are not able to allow them full free-range.

**Campbell** – reputedly the best egg-producing breed provided you get a good quality strain. Most commonly they are khaki or white in colour and they lay up to 300 medium-sized white eggs a year. They can be skitty, but are largely flightless, only able to make short runs or low heights. As they are a small to medium-sized bird, the drakes do not dress well and they are best kept just for eggs.

It is worth noting that our personal experience is that our Campbells are not as good layers as our Indian Runners. There are no guarantees when dealing with living creatures! Some strains of a breed will be better than others and individual birds will vary as well.

**Diseases**
With young ducklings Aspergillosis can occur. This is a disease usually found in a duckling that affects the lungs, presenting with breathing difficulties and usually caused by mould being inhaled by the ducklings either at hatch or from

mouldy bedding. It usually presents itself within the first 5–10 days. If a duckling suffers from obvious breathing difficulties, veterinary treatment should be sought as soon as possible. Those ducklings affected will need to be placed into a separate brooder to prevent infection of the rest of the hatch.

Wet Feather is a problem with both ducks and geese. See page 95. Ducks are also susceptible to some of the same diseases, illnesses and parasites as chickens such as worms, coccidiosis, and mites and they should be treated in the same way as described in the Disease and Problems chapter.

# 12

# TURKEYS

Over recent years the raising of a turkey for Christmas in the back garden has become more common, encouraged by the example of a few television chefs. A fair percentage of these birds have ended up as pampered pets living out their lives as family pets. When the day arrived to turn them into Christmas dinner, it just couldn't be done. Sadly, they don't live very long – three years being normal and five for an old timer.

Turkey eggs are delicious, but don't expect many. If you get one a week on average you'll be doing well. Turkeys aren't economically viable as egg birds; you're raising a table bird, and as Christmas turkeys tend to be culled by the time they are 26 weeks old, they are too young to start laying.

There is a huge difference between a home-raised turkey and a commercially raised bird that you will buy in a supermarket or from a butcher. The former is even better than the free-range birds sold at a premium. Because they have grown slowly and had a more varied diet, the meat is moister, has a firmer texture and is far better flavoured, plus you know they have been kept in the best, most humane conditions possible.

Turkeys are far more communicative and interactive than you might expect. If you are with them from hatching, they will consider you their mother and follow you everywhere. It's impossible to do any work around them in the garden; all they want is to be near you and see what you are up to. They're great fun though.

It's probably best to buy your turkey poults, as the young birds are called, when they no longer require artificial heat, described as "off heat". The reason is that turkeys are more difficult to raise than hen chicks and suffer a higher mortality rate even with the best of care.

When first hatched their eyesight is very poor. This may hinder them from finding their food and water. Their eyesight does improve after the first week but, in the beginning, in addition to a heat lamp, you may need to provide a brighter light over their food and water so they can see it.

If your heat lamp is infra red, that may give enough light. Some people use a white light, but we find that the red light is less stressful for them. If you notice one bird isn't eating, dip her beak into the feed to get her started, or pick up a few crumbs and drop them onto the rest of the food. This imitates what the mother turkey does and the poults will follow the sound of the food dropping.

For the first few days, put their feed into a shallow tray, such as a shoe box lid or egg tray. When you first get the poults home, place them on the tray. Their first reaction is to eat what is under their feet, in this case, their turkey crumbs!

On hatching, they will have absorbed the egg yolk. This is enough to keep them going for the first four or five days so you may not realize they're not eating or drinking until they die for no apparent reason. The quicker you can get them eating and drinking, the better their chances.

As mentioned on page 114, with hen chicks we put marbles in the water bowl to stop them drowning in it. We do the same with turkeys. However, if you use a specific chick drinker, you don't need to worry about them drowning. In fact, a plastic chick drinker is about the same price as a bag of marbles in some toy stores! We also put some marbles into the turkey's feed as the reflective shiny surface seems to attract them and encourage them to eat.

Turkey poults are very sensitive to cold and love the heat. The ideal temperature to keep them in is 37°C for the first week and then the temperature should be gradually reduced by a degree a week until they are fully feathered up at 5 weeks or so. Even then, do not suddenly remove the heat altogether. In cold weather or just for cold nights, provide additional heat for another couple of weeks.

As with hen chicks, you'll know if they're too cold as they'll tend to huddle under the heat lamp and they'll cheep more. Happy poults tend to be quiet. When you've got a dozen or so chicks, this huddling behaviour can result in those in the middle being smothered. Conversely, ensure they have enough room to move away from the heat lamp if they're too warm.

It's important not to give them cold water to drink. Tap water is probably at 5°C and this chills them internally. Ideally you could have two water bowls or drinkers, one in use and one warming up to at least room temperature. Or you can add a little hot water from the kettle to bring its temperature up when you refill. Keeping the bowl or drinker just outside the perimeter of the heat lamp will make sure it's not cooled down too far as well. Not directly under the lamp though; you don't want the water getting too warm.

The poults also have a tendency to eat their bedding, which is obviously not good for them. Putting newspaper down to enable easy cleaning isn't the answer either as they slip about on smooth surfaces. We found that a bottomless box on a rough concrete floor or a wooden shed floor was best. We could move the box, heat lamp, etc., to a new spot and clean up the old, which was quite a lot of work.

Friends of ours use a base of wood shavings, but for the first couple of days they cover the shavings with an old clean sheet or pillowcase. This works if you only have a couple of poults. The poults don't eat the bedding and they don't slip on the surface which they could do if you used newspaper or similar.

As you can see, brooding is quite a lot of work and whilst it is rewarding, you can see why we suggest obtaining your birds when they're off heat.

## Housing

Because they're a large bird, they need a larger house and run than a hen. A small garden shed, say 6 ft x 4 ft (1.8 m x 1.2 m) would be good for two birds but a little cramped for three. Just like chickens, they hate draughts but need ventilation. Ensure there are vents in the top of the shed but no draughts lower down to chill them.

Despite being large birds, they like to perch. A bar of 3 inches x 3 inches (7.5 cm x 7.5 cm) nominal wood planed all round with the edges chamfered is ideal, placed about 3 feet (90 cm) above the floor. A thick layer of wood shavings on the floor will serve to absorb the droppings, keeping the house sweet. A thick layer of shavings also helps to cushion them when they drop down from their perch and hit the floor.

Their bones are quite soft for the first three months so allowing the birds to perch when young can give the breasts a blister on the front and can misshapen the breast bone. You may prefer to put a perch in after three months or not at all.

Once grown, they're hardy birds and cold weather won't trouble them. They do need shelter from wind and rain though. Windy wet weather will chill them. If they're in an enclosed run, a tarpaulin over part of the top and another down the windward side will give them protection. Sometimes they're a bit silly and you need to keep them in the sheltered portion in bad weather.

The run should allow between 4 m$^2$–6 m$^2$ (43 ft$^2$–65 ft$^2$) per bird unless you are letting them have the run of the garden. Obviously you need to give the same consideration to foxes and predators as with other poultry.

Once fully grown, they are largely flightless due to their size and weight, but the younger birds can manage heights up to 6 feet (1.8 m) so this should be considered when allowing them to range freely.

Turkeys can be prone to panic. A loud noise at night can cause them to damage themselves trying to escape the shed, sometimes through the wall. This is something to be aware of on Bonfire Night, which seems to stretch across two months nowadays.

A small light in the shed at night, even a solar powered one if you've not got a mains supply to it, will help prevent panic injuries. Hopefully your birds will know you from your chatting to them when you feed, etc. So if you go into the shed at night be sure to chatter away to them as you get to the door. It doesn't matter what you say so long as your tone is calm and reassuring.

## Feeding

Turkeys grow to a much larger size than chickens so their nutritional requirements are different; they require more protein in their diet. You need to buy feeds specifically compounded for turkeys rather than chicken feed.

Start with a turkey starter crumb and after three weeks add a little sprinkling of poultry grit to the food for their gizzard. Just a pinch is sufficient, not too much. After five weeks start to move them onto turkey rearing/growers pellets. Mix 50:50 for a week and then drop off the crumb. Don't give them crushed oyster shell or layers pellets. They do not need the same level of calcium as laying hens do and it is harmful if they have too much. If you are raising turkeys from chicks, adding some vitamin supplement to the water is a good idea but not absolutely necessary.

Just like chickens, they love treats and some greens (when grown) but don't kill them with kindness. Think of the treats as giving children sweets: a few do no harm but they need to eat their dinner!

## Breeds

**Norfolk Black Turkeys** – these have lovely and striking black feathering and are slow to grow. They have a tasty full flavour with large legs and smaller breasts. They are better suited to backgarden keepers and those who wish to keep for breeding as they tend not to grow overly large so do not suffer from illnesses associated with larger breeds.

**Bronze Turkeys** – these are prized for their taste and size. They have lovely bronzed plumage and are slow-growing, giving a stronger and fuller flavoured meat.

**Hybrid Turkeys** – there are many hybrid varieties of turkey that are fast growing and more suited to commercial enterprises as opposed to backgarden keepers and breeders.

## Diseases

Generally turkeys, once grown, are fairly healthy. They do share some diseases and problems with other poultry, such as bumblefoot and coccidiosis. A medicated feed to start with will prevent coccidiosis.

There is one disease that you really need to concern yourself with – blackhead, or to give it its proper name, Histomoniasis. Chickens can suffer with this disease, but usually it doesn't bother them. However, they can pass the problem onto turkeys and for this reason it isn't a good idea to keep chickens and turkeys in mixed flocks or to keep turkeys on ground that has recently had chickens on. If you are going to keep turkeys on land that has chickens on, then cleaning the land by applying a heavy dose of lime, as much as 8 oz per square yard (250 g per square metre), and allowing it at least a month to work, will clean the land but remember nothing is absolutely certain.

The disease is caused by a parasitic worm. Chickens usually just pass them through their bodies without being affected by them. However, they will kill your turkeys. The worm attacks the turkey's intestines and liver which reduces the blood supply, resulting in a darkening of the head, hence the name blackhead.

The symptoms of the disease include lethargy, loss of condition and weight loss, as well as the darkening of the head. There is one other symptom that is very easy to spot and very indicative of the disease: sulphur yellow droppings.

Once they have the disease, there is little you can do apart from culling to prevent suffering. It's possible that a vet could prescribe an anti-microbial drug called Tiamulin in the early stages but at the time of writing there is no other potential treatment available.

Prevention is better than cure, especially when there is no cure, so regular worming every six weeks is a must if you even suspect your birds may be at risk. Keeping the area clean of droppings and maintaining a high standard of cleanliness is essential, especially within mixed flocks. As well as using lime, using a dry disinfectant such as Bio-Dri, or using a regular wash down with something such as Jeyes fluid, will help to sanitize the area and protect your birds.

**Preparing for the Table**

You should look to cull your bird some ten days before you
intend to eat it. Once the deed is done, pluck immediately and
then hang the bird. You want to ensure that no flies, etc., can
get to it. We found that popping the bird into a pillowcase tied
with string around the feet did the job.

Unless you're lucky enough to have a walk-in fridge or
coldroom, choose a cold area like an unheated shed or garage.
Usually temperatures are fairly low in December anyway but
be careful if hanging the bird at warmer times of the year to
keep it somewhere very cool.

# 13

# GEESE

We were initially unsure whether to include geese in a book about backgarden poultry. Unlike the ducks, geese are best kept in a very large garden or on a smallholding as they need large grazing areas. Most small keepers will raise a few birds for meat for Christmas or other holidays, but they can make good pets, good guard dogs and they do lay a tasty egg.

Geese have been kept and bred by man longer than any other species of poultry. There are illustrations of domestic geese in Egyptian tombs dating back some 4,500 years. Their reputation as a guard dog dates back to the sacred geese of the temple of Juno alerting the citizens of Rome to a night attack by the Gauls in 390 BC.

Despite their reputation as guard dogs, most geese are fairly docile although the ganders in particular can prove aggressive in season and as such are not ideal with children unless kept in an enclosure. All geese will raise a racket, honking at the approach of strangers or at a predatory threat which can seem very threatening. Despite what you may have heard, a gander is no match for a fox and you will need to keep that in mind when allowing them to roam.

## Housing
Geese are large, hardy birds but they require secure housing that offers shelter from the rain and wind with good

ventilation. A shed with the windows replaced with strong wire netting or weldmesh to keep out foxes is ideal.

As with ducks, they do not perch and a ramp up to a stepped door will make it easier to get them into the house at evening.

## Grazing, Feeding and Watering

Geese are primarily a grazing bird and need access to pasture to thrive and enjoy life. Long grass over 4 inches (10 cm) high can cause constipation and impacted crop so do not graze them on unkempt land. Either top the pasture (remove any long grass) or strim off tall grass lawns. Birds being raised for the table will need a minimum of 200 feet$^2$ (20 metres$^2$) each but the grass will be in pretty poor condition at the end.

When keeping geese permanently on land, you need to allow far more than 200 feet$^2$ (20 metres$^2$), perhaps ten times as much, and rotate around the area to avoid damage and the build-up of worms and parasites.

If you are fortunate enough to have an orchard, they'll love the shade in summer but they can strip the bark from young trees so you'll need to guard the trees against the geese.

Once the grass starts to grow in spring and right through to the autumn, geese will get much of their nutrition from the grass. Different grasses, or more accurately grass mixes, will provide different levels of nutrition. This will also vary depending on the land and the weather.

Supplementary feeding in the season and once the grass growth slows in the autumn will, therefore, be required. A ration of grain fed in the afternoon, as much as they can eat in 15 minutes should suffice. You can scatter this on the ground or provide it soaking in water to soften it which they tend to prefer. Feeding is pretty much the same as for ducks except, being larger, they eat more.

When buying commercial compound feeds, as with ducks, ensure you buy feed specific for waterfowl. Never feed your geese on medicated (against coccidiosis) chicken feed as such food contains a medication that is harmful for waterfowl.

As with all poultry, they must have continuous free access to clean drinking water. Like ducks, geese love to swim in a pond and the larger breeds, in particular, need a pond to enable them to mate. If you have a clean stream running through your garden, you may be able to run it into and out of a pond, the flowing water keeping it clean. Otherwise you can use the same methods as we describe with ducks to provide bathing water (see page 117).

Again as with ducks, you do not actually need to provide water they can swim in but you must provide water that they can immerse their heads in. A large bucket will suffice.

## Obtaining Stock

The easiest way is to buy in young stock which will be hardy, although you can obtain fertile eggs to incubate or young goslings to grow on. You can buy adults and breed them, usually as a trio – one gander (male) and two females. Geese are quite good at raising their young but you will need enough land for them.

## Breeds

**Embden Goose** – the largest goose and ideal to be raised for meat. They have a large white body, orange bill, blue eyes and can reach nearly a metre in height in good strains. The Embden is not an egg producer, with only 12–20 eggs a year in season but a dressed bird can weigh up to 20 lb (9 kg) which more than makes up for its poor laying habits. Due to their size, they require large grazing areas.

**Chinese Goose** – a slightly strange-looking bird with a large knob on the crest of the beak which is larger and more pronounced in the gander than the goose. They are likely to lay up to 60 eggs in season and both the goose and gander have a good-sized lean carcass for meat production. Chinese Geese are smaller than Embden or Toulouse Geese, and come in white and brown/grey colours and have blue eyes with an orange bill. Although they can fly, they tend to stick to an area but can be pinioned or wing clipped to keep in a set space.

**Toulouse Goose** – another great-looking bird with grey feathering, brown eyes and a bright orange bill. They are the most gentle of the different geese breeds and can also be kept in smaller areas but they do need a proper pond due to their size. If crossed with the Embden, the offspring are both quick-growing and larger than their parents and ideal for quick meat rearing.

**Sebastopol** – these originated in the Balkans and were imported into Britain in the mid 1800s. At that time the accepted English spelling was Sebastopol but nowadays the city is spelt Sevastopol but for the goose the "b" remains in the name. The first reference we could find to them was in *The Poultry Book* by W. B. Tegetmeier, published in 1867, which states, "*Amongst the geese there were two curious specimens from Sebastopol, exhibited by Mr. T. H. D. Bayly. These birds are somewhat smaller than those of this country at a mature size, but they are of the purest white and the most perfect form, whilst the most conspicuous portion of their plumage is of a curly nature, affording a very striking contrast to the feathers of the ordinary English goose.*"

# 14

# QUAIL

For those with very small gardens or who are interested in raising a different species to produce a variety of eggs or meat, quails are a lovely addition to the back garden flock. Because they are small birds and take so little room, you could even keep a few quail on a balcony if you didn't have a garden.

Raising and keeping quail is not an expensive hobby. In fact, you can often make some profit by selling the eggs, which they lay in abundance. Although they're very small, the eggs fetch a premium price as a gourmet delicacy. Many small-scale poultry keepers are expanding into this area to sell the eggs and the quail as meat birds.

They make a good pet, being fairly easy to tame, especially if you've raised and handled them from the day they hatch. Even so they can be skittish, but patience and slow movements will win them around.

Unlike the other poultry discussed, quail are actually game birds rather than poultry but that doesn't stop them being prolific egg-layers, producing as many as 300 a year. Quail are only hand-sized and have a life span of around two years, but in most breeds the females will lay consistently from around 6 weeks until 18 months of age so they are a very productive choice for the urban poultry keeper.

You will discover that your quail make a lovely set of chirrups to one another during the day. The only problem being that at dawn the males can often make a loud mating call which if kept close to the house can be very noisy, so make sure that this doesn't bother you or any neighbours by placing the aviary a little way down the garden away from the house if possible.

The males, if being raised for meat, are ready by 8 weeks of age so are a quick bird to raise for the table. It is a good idea to separate the males from the females at 4 weeks of age to avoid aggression in your flock. We have found that as meat birds they have a lovely taste, slightly gamey but with soft meat. One quail is perfect as a starter or two as a main course for one person.

Quail are best kept as a quartet with one male to three females. Over stocking, too high male:female ratios and lack of interest provided in the coop or aviary will lead to boredom. Bored quail often peck at each other and can inflict serious injury or death. To avoid this, we add a few dangling CDs, a peck block and give them fresh greens.

The average lifespan of a domestic quail is 3–4 years with egg-laying dropping off in the third year. The oldest recorded quail lived for ten years.

## Housing

Quail are very easily kept in the garden in either an integrated coop and run set-up or in a converted rabbit hutch or an aviary. You need to make sure that the mesh or hutch bars are small enough so that the birds cannot escape. Their ability to fly should be kept in mind when cleaning out cages, etc. Just one moment's inattention and your bird can be flying away, never to be seen again.

We would recommend no more than one bird per 1.5 ft$^2$ (0.14 m$^2$) of overall floor space as they are permanently resident in the area and aggression is a common problem with overcrowded quail. One problem with quail is their habit of flying straight up when startled. This can result in them knocking themselves out against the top of the run. If you have an aviary with a fox-proof welded mesh top, you might need to put a softer plastic net a few inches below as a collision barrier.

The housing should be cleaned out regularly, at least on a weekly basis, with fresh bedding and litter material so that no sores or infections can take hold in the birds.

For feeding and drinking, the smaller chicken feeders are good and will provide plenty of food and drink for ad lib feeding. We like to use a chick feeder as we find they mess in it less than in the normal chicken feeders. This is a common problem with quail. They also often mess in their water so you will need to change the water and clean the drinker on a daily basis to keep it clean and avoid any disease or illness in your birds.

We keep a quartet of quail in a double rabbit hutch with housing and nesting areas above and a run below with a ramp down to the lower level. As with our other poultry we use Hemcore in the house as well as in the quail's run. This is long-lasting and they enjoy using the deep litter in the run for dust bathing.

Quail are not as hardy as poultry and their ideal temperature range is 16–23°C. Providing extra bedding, clean straw, etc., for them in winter helps them to keep their temperature up. In cold weather, you will probably need to provide heating in the coop or even move them indoors into cages in extreme conditions.

## Obtaining Quail and Incubation

Cotunix quail can be purchased easily either as eggs, day-old birds or at point of lay from specialist breeders.

Remember that when buying incubating eggs, you don't have any control of the sex of the birds. Usually this is half and half, male to female, but you can be very unlucky and find that you've got six males from six eggs.

In the absence of a broody quail, you can use a broody bantam. Bantams may well hatch the eggs for you better than a quail. The average hatching time for quail eggs is 16 days although this varies according to breed, as opposed to 21 days for chickens and 28 days for ducks.

If you are using a mechanical incubator, you should incubate to give an egg temperature of 37.5°C. Refer to the individual machine's instructions. The eggs should be turned daily (pointy end down). Make sure you maintain humidity as per your incubator's instruction manual. Ideally 45 per cent until they start pipping, rising to 75 per cent until hatched. Once the quail start pipping, cease turning and reduce the temperature to 37°C.

Once hatched, dried and fluffed up, move them to a brooding area with a heat lamp. This need not be anything more sophisticated than a cardboard box with newspaper and wood shavings on the floor. They need to be kept at 35°C for their first week and then reduce their temperature by 2.5–3°C each week until at 6 weeks they are totally off-heat.

Provide water in shallow dishes. A small pet food bowl is fine, but put some large pebbles or glass marbles in the water to make sure the level isn't too deep. Quail are susceptible to drowning in the water.

### Rearing

Quail should be fed on either a game bird starter mix or if you also rear chickens an un-medicated chick crumb feed for the first two weeks.

After this, you should begin to move the quail onto growers feed until 6 weeks of age, at which point the birds will be physically mature. For the male birds destined for the table, follow with one week of finishers feed. For the female birds, move them onto layers pellets or mash at this point. Using the finishers pellets will help the final table bird to be a good body weight and also helps to improve its flavour.

A personal choice for flavour would be to give 50 per cent corn along with the finishers pellets to give a well-flavoured and succulent bird.

### Diseases

Quail can suffer from many of the same diseases, illnesses and parasites as chickens, such as worms, coccidiosis, and mites. They should be treated in the same way as described in the Diseases and Problems chapter.

Although not a disease, feather pecking can be a particular problem with quail. Quails were actually kept as fighting birds in Elizabethan times!

Even when you keep the male:female ratio right and give them plenty of space, feed and entertainment, you can find that pecking occurs. We find that the purple antiseptic equine spray is invaluable and have noticed we use this far more with the quail than the chickens. Separating a bully for a few days and making sure injuries are quickly treated and covered will help, as at the sight of blood quail can quickly further injure or even kill one another.

In the spring months, it is a good idea to keep an eye on any males to make sure they are not pulling head feathers out during mating. If they are, spray the area with purple spray so that the others do not catch sight of the exposed skin on the head.

**Breeds**
There are many different breeds of quail. We only have space to list the most common here.

**Japanese or Coturnix** – a very hardy little bird that is usually mottled brown in colour, with the females having a speckled chest, although other breed variations of colour are available. They start laying from 6 weeks of age and will lay almost daily unless disturbed.

**Bobwhite** – these most commonly have a brown buff colour, with the males also having a white throat and eye stripe. The hens will lay in a clutch and will try to sit unless they are regularly removed.

**Californian** – an unusual-looking little bird with a brown-feathered forward-hanging helmet, grey and brown feathers and a black head, with males having white facial markings. As with the Bobwhite, the hens will lay in a clutch and will try to sit unless they are regularly removed.

**Jumbo** – as the name suggests, a large quail essentially bred as a commercial table bird.

**Chinese Painted Quail** – a compact breed more ornamental than utility. The males can be spectacular.

# 15

# EGGS

*We hear a good deal said in these days about the 200 egg hen.*
*Some are disposed to deny her existence, and to class her with*
*such fabulous or semi-fabulous birds as the phoenix and*
*dodo. Others admit that she has appeared in isolated*
*instances, but is by no means common. Others contend that if*
*she should appear in large numbers it would be a misfortune*
*rather than otherwise, for such excessive egg production*
*would weaken her system so that her eggs would not hatch*
*healthy and vigorous chicks; and the 200 egg hen would be in*
*constant danger of extinction from her own success.*

*One thing is certain, however, the 200 egg hen is no myth.*
*There are many of them scattered about, and the tribe is on*
*the increase.*

From: *200 Eggs a Year Per Hen: How to Get Them*
by Edgar Warren, 1912

The reason most people start keeping poultry is quite simply
the joy of fresh eggs. We've never found an egg in a shop,
even organic and free-range, that is quite as fresh and tasty as
one that our hens have laid.

Of course, once you have your birds the fun of keeping
them as pets or the enjoyment of breeding and showing them
takes over and the eggs just become part of the experience.

The egg really is a miraculous thing. Nature has developed
a packaging that balances the protection of the egg and

eventual chick with the necessity of allowing the chick to break out at the end into the wide world. In one container is all that the chick needs to develop, from a couple of cells up to a creature able to walk and feed. Just add warmth for a fluffy chick!

Of course, the vast majority of eggs we eat are not fertilized and so will not produce a chick at the end. However, nature's efforts to protect and produce the next generation have resulted in a food that is not only nutritious and tasty but also designed to store well.

It's worth restating: you do not need a male bird, cockerel or drake to produce eggs for eating. This is the most common myth about poultry keeping. A cockerel does not encourage the hen to produce more eggs. The number of eggs produced depends on factors such as day length and the age of the bird.

## How the Egg is Made
The production of an egg within the hen takes slightly over 24 hours from ovulation to being laid. It starts with the ovum being released and moving to the infundibulum, the funnel shaped organ at the top of the oviduct. Here it is fertilized if sperm are present but if no sperm are there it still follows the same process.

After mating, the hen can store the cockerel's sperm for about three weeks. Incidentally, it takes about three days from mating for the sperm to get to the infundibulum so if you are breeding and looking for fertile eggs, it's best to wait a week after mating before incubating.

The yolk is the first thing to form and then it is covered with the vitelline membrane followed by structural fibres and the albumen (the egg white). The final part of the process is the formation of the eggshell (made of calcite – calcium carbonate) which is deposited around the egg in the lower part of the oviduct right before it is laid.

## Colour of Eggs
The colour of the egg is due to pigment deposited on the shell towards the end of its formation inside the hen. Like paint, it's

| Breed | Egg Numbers Per Year | Usual Age of Lay | Egg Colour |
|---|---|---|---|
| Araucana | 180 | 20–22 weeks | Blue to Blue/Green |
| Buff Orpington | 220 | 26–30 weeks | Brown |
| Cream Legbar | 200 | 26–30 weeks | Blue/Green tinted |
| Gold Brahma | 200 | 26–30 weeks | Light Brown |
| Light Sussex | 240 | 26–30 weeks | Creamy-Light Brown |
| Maran | 220 | 26–30 weeks | Dark Brown |
| Rhode Island Red | 220 | 26–30 weeks | Brown |
| Silkie | 160 | 30–34 weeks | White |
| Warren | 280 | 20–22 weeks | Brown |
| Welsummer | 220 | 26–30 weeks | Brown |
| Wyandotte | 200 | 26–30 weeks | Brown |

slightly wet and can be smeared immediately on an egg being laid.

The colour is purely down to the breed of hen that lays the egg. In the UK, eggs were generally shades of brown until the post-war period when mass production arrived. The breeds that were best for egg-laying in battery cages generally laid white eggs and so the marketers implied that white eggs were somehow more pure than brown eggs, which is obviously nonsense.

When people started to be concerned about battery eggs, they related the colour to the method of production and so there grew a common belief that white eggs came from hens kept in battery cages and brown eggs came from free-range hens.

Of course, the marketing men aren't stupid so they changed

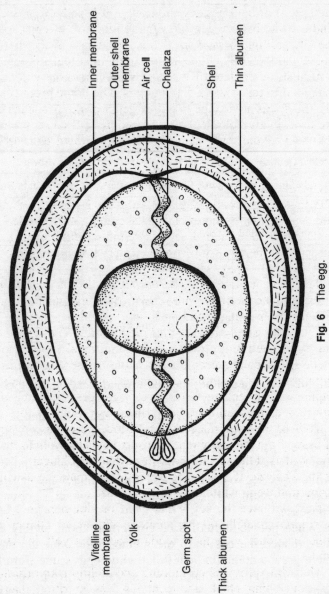

Inner membrane
Outer shell membrane
Air cell
Chalaza
Shell
Thin albumen

Vitelline membrane
Yolk
Germ spot
Thick albumen

**Fig. 6**  The egg.

the breeds of battery hens and now you'll be hard pressed to find a white battery egg. By far the majority of eggs sold in the UK are light brown but this isn't the case everywhere. In Spain and the USA, white is the common colour for eggs.

Whilst the content of the egg is the same, the colour can make a difference when selling eggs. The home producer can use this to his advantage to gain a better price. Araucana eggs are a distinctive blue colour and people will pay a premium for them as they will also for the dark brown eggs from a Maran.

## Structure of the Egg

The first thing you'll notice is that the egg has a fat rounded end and a pointed cone end. The egg is laid rounded end first. Under the rounded end is the air sac that forms as the contents of the egg cool from the internal temperature of the bird.

The shell itself is slightly porous and allows air to pass through to the interior, whilst still providing protection for the contents from damage and bacteria.

Under the shell are two thin membranes. The first is just under the shell and provides some additional support to the shell. The second surrounds the white of the egg and as the egg ages the two membranes separate. The fact there are two membranes and that they split as the egg ages is why it is difficult to peel the shell from boiled fresh eggs.

To hold the yolk in a central position, there is a connection of tissue from the membrane that surrounds the yolk to the top and bottom of the second membrane that is technically known as the chalaza. If you crack an egg, you may see a whitish string in the egg white; this is the chalaza.

Next we have the egg white itself or albumen. In a hen's egg approximately 60 per cent is white and 40 per cent yolk. In a duck egg, there tends to be more yolk in ratio to white.

The white provides cushioning and additional nutrition for the developing chick and contains less nutrition than the yolk.

There's another membrane surrounding the yolk and then

the yolk itself. On the yolk you may see a little white spot. This is the germinal disc which contains the cells that will develop into a chick if the egg is fertile.

## Nutritional Content of the Egg

Eggs are an extremely rich food on a par with meat for protein content. They also contain a wide range of vitamins and a certain amount of fat. From being described as a near perfect food, in recent times a concern arose that eating eggs would increase your cholesterol levels and that they were, therefore, bad for you. Then further research from the University of Surrey suggested that most people could eat as many eggs as they wanted without damaging their health. Accordingly the British Heart Foundation dropped its advice to limit egg consumption to three a week in 2007 in light of this new evidence. Only 27 per cent of the fat in an egg is actually the harmful saturated fat that is implicated in heart disease and excess cholesterol.

Another myth about eggs is that they cause constipation – some confusion about the phrase egg-bound probably being the cause of this. Eggs do lack fibre which is necessary for our bowel but eating eggs is not in itself the cause of the problem.

A large egg (around 2 oz/50 g weight) will contain around 75 calories. An average daily requirement for women is around 2,000 calories, so a couple of eggs a day won't make you gain weight. The egg is approximately 10 per cent fat and 12 per cent protein but only 1 per cent carbo-hydrates.

On the subject of slimming, there was a fad diet some years ago based on eating as many hard-boiled eggs as you wished to fill you up. We suspect that depends on how much you like hard-boiled eggs!

Most of the nutrition in an egg is in the yolk. The white typically contains a quarter of the calories of the yolk, despite the white being 60 per cent of the egg. Chicken eggs supply all the essential amino acids we need plus vitamins and minerals, notably:

Vitamin A
Vitamin B1 (Thiamine)
Vitamin B2 (Riboflavin)
Vitamin B5 (Pantothenic acid)
Vitamin B9 (Folic Acid)
Calcium
Iron
Phosphorus

The one big deficiency of eggs as a food is that they don't provide any vitamin C or fibre. Personally, we'd suggest you eat as many eggs as you want as part of a balanced diet.

There is some evidence that free-range eggs are better for you than caged eggs since the exact composition of the egg will depend on what the hen has eaten. The more varied diet of the free-ranger tends to help in producing eggs with an even higher ratio of unsaturated to saturated fats and higher in omega 3 fatty acids

Recently eggs that are especially high in these omega 3 fats have come onto the market. The birds are fed on special diets high in polyunsaturated fat which benefit the eggs. Since a free-range egg is better for you anyway we see little real benefit in this idea but "you pays your money and takes your choice".

One area of real concern is chemical residue in shop bought eggs. The Soil Association discovered that many eggs in the shops contained residues above the safety level of various chemicals including pesticides. Most worrying was the discovery that some of these residues were of chemicals not even approved for use on laying hens.

With home kept poultry, you are in control and can guarantee that your eggs do not contain residues by only using worming products, etc., in strict accordance with the rules.

**Eggs from Different Species**
Commercially there are three species of poultry eggs commonly available: hens, ducks and quail. Chicken eggs are by far the most popular, with duck eggs being more likely to be found in independent shops and market stalls than in the

supermarket. Quail eggs are often found in delicatessens as they're considered a delicacy.

Duck eggs do have a slightly stronger taste than hens' eggs and a different consistency; slightly chewy is the best description. They create a really rich scrambled egg and are wonderful for baking.

Quail eggs are very small and quite bland tasting. Most often they're served or used in dishes having been hard-boiled. Peeling them can be very time consuming. Boiling in a mixture of half and half vinegar and water makes peeling easier.

The reason that commercial egg production for eating is limited to the above-mentioned three species is simple economics. They are the species that have breeds that lay prolifically. Under 200 eggs a year it becomes near impossible to produce an egg that people will actually pay a price for that covers the costs of production. The cost of looking after a bird is more or less the same whether it lays 50 or 300 eggs a year.

It may well be possible with selective breeding to increase the number of eggs laid by geese and turkeys but there just isn't the market to justify the effort.

Goose eggs are delicious, quite large and mostly white. Sadly you'll be lucky to see 30 eggs a year per bird. Turkeys produce slightly more eggs than geese, up to 70 a year with a breed like the Norfolk Black, but again not enough to be commercially viable.

**Double Yolk Eggs**
Most of us have had the experience of finding a double yolk egg. Depending on whether you are a yolk person or a white person it's either a bonus or a fault. Personally we think it's a bonus, but we prefer the egg yolk to the white.

Double yolk eggs are actually fairly rare, about 1 in 1,000 for commercial eggs where consistency is required and has been bred for. Rarer still, but not impossible, are multiple yolk eggs, triples or even quadruples and we have even heard of a nine yolk egg!

When an egg starts its journey inside the hen, the ovum in

the hen's ovary is formed first. This grows and the colour changes from pale grey to the yellow we know as the yolk colour.

Once it reaches full size, the yolk sac breaks away (ovulation) and begins a journey down the oviduct where the egg white (albumen) and the shell form around it. Normally, the next ovulation is triggered by the hen laying the egg but occasionally things go wrong and two yolks are released at the same time to travel down the oviduct together, being surrounded by one shell and giving us the double yolk egg.

If the ova are fertile, the double yolk egg will contain two viable chick embryos but there will not be enough space for them to develop to hatching. Unfortunately twins from the same egg are not really possible.

Most often double yolk eggs are laid by young hens of productive egg-laying breeds. If you really like double yolk eggs, then the highly productive breeds are more likely to reward you when young.

As the hens become more mature and their system settles down to correct production, the double yolks become less frequent to non-existent.

Since the double yolk egg cannot bring forth double chicks, genetically it is not possible to have a breed that consistently produces them. They'd die out! It is possible in theory to develop a breed where the ratio of double to single yolk eggs is higher but we do not know of one.

Hens aren't the only bird to lay doubles – ducks do it quite frequently as well. In fact, any bird can lay double yolkers but it is relatively more common in birds bred to lay rather than wild birds where genetic selection works hard against it.

So if you get some double yolk eggs from your hens, make the most of your good fortune because it won't last.

## Egg Faults

On occasion you will find faults with your eggs, and it's useful to know the causes, even though not all problems can be prevented.

*Wind Eggs*
A wind egg is an egg without a yolk. They're often misshapen as well as not containing a yolk. Usually this is a problem associated with birds starting to lay for the first time and is nothing to worry about. Once the bird settles into routine, the problem goes away. It can reoccur though towards the end of the bird's laying life.

*Pale Egg Yolk Colour*
The yolk colour of eggs naturally varies from light yellow to a deep orange yellow depending on what the hens are eating. Grass and other plants, such as clovers and lucerne, enhance the yellow colour of the yolk. Some corn will help darken the yolk and tomatoes have an effect.

Commercial egg producers have been known to use various additives to the bird's diet to achieve a deeper colour in the yolk, simulating the colour of yolk of free-ranging birds with a varied diet.

Ill health can also affect the colour of the yolk. If one of your hens starts to produce eggs with a much lighter colour than others in the flock, it could indicate an underlying health problem.

*Green Egg Yolk Colour*
Sometimes free-ranging hens will produce eggs with green yolks, most frequently in the spring when plants are most lush. The birds should be given more compound feed to reduce the proportion of greens in the diet to correct the problem.

*Rotten Eggs*
Usually the shell membranes that lie just under the shell protect the egg from microbial and fungal infection. If this layer is damaged or malformed and infection gets into the egg, rot in patches or whole will occur.

Sometimes you will find a clutch of eggs in some obscure place in the garden and think them freshly laid. It's best to keep them separately and crack into a cup before using in case they are older than you think.

## Bubbles in the White

Eggs normally have an air space at the blunt end and the shell is permeable to the air to allow oxygen in to developing chicks. If, however, the inner membrane is damaged, the result can be bubbles in the white.

## Blood Spots in Eggs

These are small red to reddish brown spots found in or around the yolk. They are usually caused by one of the tiny blood vessels in the ovary breaking at the time when the yolk is released. Often people mistakenly think they indicate a fertile egg and it is the start of a chick forming.

High levels of activity or disturbance, particularly at the time of ovulation, are likely to increase the incidence of these blood spots.

Because free-ranging hens may eat grass which contains a substance called rutin that has the effect of stopping bleeding, free-range hens' eggs tend to have fewer blood spots than those from caged or battery kept birds.

## Meat Spots in Eggs

Meat spots are usually brown in colour, darker than blood spots, and they are found in the egg white (albumen) rather than the yolk. They consist of small pieces of body tissue, such as the internal wall of the oviduct. Their incidence varies according to bird age and health and also due to breed differences.

In brown shelled eggs, they are more difficult to identify when candling and brown egg-laying hens are more likely to produce them than white egg-laying hens.

## Watery White (Albumen)

Although eggs will remain safe to eat for at least 28 days, the internal quality begins to deteriorate from the day they are laid. Newer eggs have firm whites that hold their shape. Poor quality whites usually indicate that the egg has been laid for longer than you realize. It's quite amazing how often you come across eggs laid where they shouldn't be some time ago.

Some newly laid eggs from older birds may have poor
quality whites and eggs. Viral disease can also cause the hen
to lay eggs with poor quality whites.

*Soft Shells*
Sometimes you may find that your hen lays you an egg that
isn't fully formed or that has a very soft shell. This is usually
a problem with the final part of the egg production during
which the eggshell is deposited around the egg. As the shell is
made up of a crystalline form of calcium carbonate, one of the
main causes of soft-shelled eggs is a calcium deficiency in the
hen.

You should make sure that the hen has access to ad lib
layers pellets as well as grit, crushed oyster shell and fresh
greens. The oyster shell contains calcium to aid in the
production of the eggshell and the fresh greens enable
the hen to absorb the calcium better, thereby further
assisting in the process while the grit helps with the
digestive process.

Other causes of soft eggshells include worms, a sudden
shock that disrupts the shell production and old age.

**Eggonomics**
OK, we admit to making up the word. On our website we
come across people who think that keeping poultry at home
could be a way to make a living. It's true that you can sell your
surplus eggs and make some money to set against the cost of
keeping them but please don't believe that you can make a
living from backgarden poultry.

The poultry industry, like most of the food industry in the
UK, is extremely efficient and produces both eggs and table
birds at extremely low cost to the consumer. To compete with
the professionals supplying supermarkets, you need economies
of scale and your neighbours are likely to take a dim view of
50,000 hens in your garden!

Proper commercial poultry or egg production is outside the
scope of this book but let's have a look at the economics of
egg production on a small scale at home. The actual
economics of your home egg production will depend on a

number of variables. We've tried to state these in such a way that you can calculate the value to yourself, without getting tied up in accounting concepts.

First of all, we're going to need housing and some equipment like a feeder and water dispenser.

Taking a guess at a cost all in of £500 and a lifespan of 10 years, that gives a cost of £50 a year to house our hens. Building your own house or converting an existing shed could reduce this figure, as could buying second hand housing, although you would be lucky to find second hand poultry housing in good condition. You could pay much more, however. Many of the cheaper Chinese poultry houses being sold now are unlikely to last 10 years and there is no allowance for maintenance or for fox-proof fencing in that figure.

Our research indicates that this is a reasonable price to house up to six hens in a secure back garden with no fox risk. Obviously housing more hens could mean that your cost per hen falls, but six laying hens are more than enough hens to provide eggs for the average family. Fewer hens and the cost increases per bird.

Next we have the cost of the hens themselves. This could vary from £1 for an ex-battery bird to £25 for a pure breed. Show quality and fancy birds will cost more but let's work on £10 and a productive lifespan of 3 years. Ex-battery hens have already been worked hard for a year and your losses due to health problems may be higher. Having said that, there is satisfaction in giving them a good life after their time in the battery and, in hard economic terms, they may well be the best bet for the home producer.

However, unlike a commercial producer you are unlikely to cull your birds once they become unproductive. You may think when you start that you'll do so, but after three years when Bluebell looks at you appealingly for a treat, what are you going to do?

For sake of argument, we'll assume no losses or veterinary costs due to illness, Mr Fox or what-have-you and ignore the awkward question of the retirement home for aged friends.

Our next cost is that of cleaning and maintenance, red mite

control and cleaning fluids, etc. Luckily they're not too expensive so £35 a year should suffice, allowing a little for repairs to the housing and unforeseen items.

The final cost is food, supplements and medication. Even free-range chickens require bought-in food if they are to maintain condition and produce all those eggs for us. This is yet another variable in the equation. Reducing the quantity of balanced bought-in rations may well reduce the quantity of eggs produced and affect the health of your chickens.

In the last few years we've seen the cost of feed increase considerably, driven by international commodity markets but layers pellets are still quite inexpensive. Currently a reasonable estimate would be £30 per year per bird but that price is variable.

We're now in a position to calculate the cost of keeping our hens. This will vary as we've said above but to illustrate we've produced two costings:

| 1. Annual Costs for Keeping 3 Laying Hens | | |
|---|---|---|
| *Housing* | £16.66 | (£50 per year split between three hens) |
| *Purchase of Stock* | £3.33 | (£10 per hen over three years) |
| *Cleaning & Maintenance* | £11.66 | (£35 per year split between three hens) |
| *Feed & Medication* | £30.00 | (Per hen per year) |
| **Total Cost per hen per year** | £61.65 | |

| 2. Annual Costs for Keeping 6 Laying Hens | | |
|---|---|---|
| *Housing* | £8.34 | (£50 per year split between six hens) |
| *Purchase of Stock* | £3.33 | (£10 per hen over three years) |
| *Cleaning & Maintenance* | £5.84 | (£35 per year split between six hens) |
| *Feed & Medication* | £30.00 | (Per hen per year) |
| **Total Cost per hen per year** | £47.51 | |

You can see from this that increasing the number of hens reduces the cost per hen and you can pop your own figures into the equation to calculate your costs. We've been pessimistic on costs. It's better to expect to pay more and save a little than the other way around.

The other side of the equation is, of course, the returns. The main factors here are how many eggs your birds lay and what price you can get for them. For sake of the argument, we've ignored the eggs you eat yourself.

One breed will lay more or fewer eggs than another so once again we're going to have to make a guess. Hybrid birds lay well but their productive life is short. Pure breeds lay fewer eggs per year but they have a longer productive life.

The hen naturally lays fewer in the winter than the summer and this effect can be mitigated by the provision of artificial light to control day length for the birds. Of course, this is an additional cost and, in any case, you may feel you are pushing the birds too hard against their nature.

Depending on the breed and how they are kept, we think you could reasonably expect between 200 and 250 eggs a year per bird. With some breeds 280 a year can be produced and some modern hybrids go as high as 350 eggs a year.

The price people expect to pay is led by the supermarkets. They are currently selling free-range and organic eggs at prices ranging from 25p to 31p an egg. This varies according to the season, less in summer when there may be a glut and more in winter when hens naturally kept don't lay so many.

The price you get for your eggs will also be affected by local competition. In country areas where a lot of people are keeping their own hens and "eggs for sale" signs are common, the price will be lower than in the towns and cities.

You will, we hope, be keeping your hens in conditions that are much better than those required to legally describe your eggs as free-range. However, you need to be careful in your description as you may not actually meet those standards and this can drop the value in the buyer's eye.

Taking the lowest shop price and lowest production level would give a cash return of 200 x £0.25 per hen per year, £50. With the highest shop price and production level, we would look at 350 x £0.34 per hen per year, £119.

From these figures you can see that there isn't a fortune to be made selling hens' eggs from the door. Of course, we could keep more chickens in the same house, reducing the cost, and buy cheaper chickens and food to increase the "profit". However, once we factor in labour we cannot make any high profits on a few hens in the garden. Besides, if you are putting production beyond welfare, then we hope you find some way to make money that doesn't involve living creatures.

We can say that our hobby has – at worst – not really cost anything much and we have the benefit of knowing that our hens are well kept and happy, as well as our eating eggs that are nutritionally better and taste better.

There are opportunities to make a bit more money from your hobby though. Duck eggs tend to fetch a premium and are not commonly found in smaller supermarkets but that is balanced by the fact that ducks are less productive than hens.

Quail eggs sell for less per egg, 19p each currently in a major supermarket chain, but quail cost a lot less to feed and house. So if there is a demand in your locality, then you can make a few pounds profit here. Quail eggs, being a bit of a delicacy item, are not universally popular though. What sells in London may well not sell in other places.

There are a couple of ways to add extra value to your eggs. As we said above, people will pay a premium for unusually coloured eggs such as blue or dark brown. Whilst we don't believe there is any difference in the taste of those eggs from that of the normal brown or white egg, we've known people who swear they taste better. Who are we to disagree?

Another way to add value to your eggs is to sell them as part of a product: pickled eggs, which are very easy to make (see page 168); or even a product such as lemon curd that uses a lot of eggs.

**How to Sell Your Eggs**
If you're going to sell your surplus eggs then it's useful to know about the eggs sold in shops. Your customers may well ask you – after all, you're now their expert!

When we go into a supermarket and buy our eggs, we're greeted with quite a range of names – Eggs from Caged Birds, Free Range Eggs, Barn Eggs, etc. But what do those names really mean?

One term that really annoys us: Farm Fresh Eggs. If that makes you think of hens happily scratching around the farmyard while a ruddy faced, rustic farmer's wife collects some eggs from a ramshackle hen house, think again. A farm, in this context, means a huge pre-fabricated shed – an industrial unit. "Fresh" just means less than 21 days old. Aren't those marketing people clever?

Usually the description is applied to eggs from caged hens. Over 60 per cent of all eggs consumed in the UK still come from battery units. Often you eat battery eggs without knowing it. For example, most mayonnaise is made using battery eggs. Whenever you buy a product containing eggs, unless it specifically states free-range you can be sure the eggs are the cheapest from caged hens.

The misshaped eggs, etc., that won't meet grading standards for retail sale end up in the catering industry – often being pasteurized and processed to increase storage life.

Caged hens live their entire short and wretched lives on a wire-mesh floor in racks with space per hen roughly equivalent to a piece of A4 paper – technically 550 cm$^2$ per bird in cages installed prior to 2003 but since then enriched cages are provided. These provide a minimum of 750 cm$^2$ per bird, along with a nest, perching space of 15 cm/bird and a scratching area. Hardly luxury.

If you have ever seen rescued ex-battery hens or visited a battery hen house, then we don't have to convince you of the evil of the system.

"Barn Eggs" is another of those misleading terms the marketing people love. It conjures up an image of those barns of yesteryear with some hens wandering around, safe from the weather and fox. You might notice that there's a trend here. Traditional equals good in food marketing. Sadly, the reality isn't quite so wonderful.

Our "barn" is yet another industrial unit, housing thousands of birds. However, the hens do have the benefit of different levels, perching space and nesting boxes: one nest box per 7 hens, crammed in at 9 hens per 1 m$^2$. One and three quarters of a sheet of A4 paper per hen. Hardly luxury is it? Better than caged, certainly, but not what most people would expect from the description.

With free-range eggs you'd think you were doing the right thing. The phrase makes us think of hens roaming around the field and farmyard, shepherded by a cockerel in small groups and living life to the full.

Sorry, wrong again. Once again we're talking about an industrial unit house but, and this is lots better, with access to a fenced-off area outside. Although the stocking density is limited to 2,500 birds per hectare (that's 4 m$^2$ per hen), most birds will hang around the shed if they do go out.

Hens are flocking birds. They tend to stick together as you'll have seen if you watch hens in a garden. This means that the area around the shed becomes stripped of vegetation and a dustbowl. Our free-range hens have the benefit in principle but the sheer numbers and their nature mean they really can't take much advantage of their undoubtedly better conditions.

Some authorities say that hens can recognize up to fifty

other hens and establish a pecking order but in these giant flocks this is impossible and the sheer numbers stress the birds and tend to cause social problems.

Finally, we have organic eggs. These are simply those from hens kept in a free-range system but fed only on organic food, ranging on organic land and which must not be fed growth-promoting antibiotics.

Whatever type of eggs you buy from a shop, you'll also find a "best before" date on the box and the packing station number. On the eggs themselves you will probably find a lion mark stamp. This means the eggs are UK-produced from hens vaccinated against salmonella.

There is also a code printed on the eggs that gives information about them. First a number, then a country code such as UK and finally a farm identification number. The first number tells you how the hens were kept: 0 for Organic, 1 for Free Range, 2 for Barn Eggs and 3 for Caged Birds.

## Selling Your Eggs from Home: the Legalities

Happily, selling your surplus eggs from home isn't subject to masses of legislation. A remarkable application of common sense from government! There are a few rules that you need to follow though. Technically selling eggs from home is known as "farm gate sales" but from a legal point of view your front door is the farm gate.

The first rule you need to follow is that the eggs must be ungraded. They cannot be split into sizes or quality grades, although you might choose to keep the largest and smallest for yourself and just offer the rest. Do remember that people buying eggs from the farm gate may well like different sizes in one box. It can be very useful when cooking to be able to use a small egg or a large one, depending on the recipe.

Obviously the eggs should be fit to eat and good quality or you might find them coming back on your windows! They should be clean, but not washed. Most buyers of farm eggs will not be too bothered by a spot of poo on the eggshell but technically you should reserve those for your own use.

The eggshells should not be damaged or cracked. A cracked

egg has its protection against microbes damaged and should be reserved for your own use and used as quickly as possible, preferably within a day or two of laying.

You need to display a "best before date" to comply with the law. Usually eggs are expected to last up to four weeks from the date of laying and you should (hopefully) be selling them within a day or two of lay. We recommend that you keep your excess eggs in boxes, and use a system to date order them so you never accidentally sell old eggs.

To comply with the law, displaying a sign saying "*Best Before Three Weeks from Date of Purchase*" will keep you legal.

If you sell your eggs by any other method, such as a local market or to a shop, even as a backyard hen keeper, other legal rules apply. First of all, you will need a *producer number*. These are available from the DEFRA Egg Marketing Inspectorate for free. This number also indicates your production type, such as organic, free-range, etc.

Don't forget that there are legal definitions and rules about using terms such as "free-range". If you've only got three hens strolling around a garden, it doesn't matter. To be "free-range" you must be able to demonstrate a stocking density below 1 hen per 1 m$^2$ (43 ft$^2$).

Eggs must be individually marked with your producer code number. You can buy cheap hand stamps and you can even write the number on if your egg sales are low and it's not worth the cost of the stamp. *The ink you use must be food safe.*

The eggs must be sold in new boxes. You cannot re-use old boxes, no matter how environmentally sound this is. If you have your eggs on a large tray, customers can bring their own old egg boxes along to pack them in but you must not keep a stock of old boxes and offer them to customers.

And, finally, you are not allowed to sell to catering establishments. Even if your eggs are wonderful, it's against the law.

It's really not worth selling surplus eggs apart from "farm gate sales" unless you have a fair amount of land and are operating as a business.

## Marketing Your Eggs

Even when selling a few eggs from the back door, it's worth thinking a little about how you market them. Please don't worry – we're not suggesting you hire an advertising agency and survey focus groups but you do need to tell people what you're selling.

Put a sign up. Make the sign large enough to be easily read from the road and make sure it's positioned so it can be easily seen. Remember that a car may be doing 30mph so it needs to be large enough to be read in the time it whizzes by. Not a bill board though or you'll have someone from the council knocking on your door.

If you find sign painting daunting, print your sign on paper using your computer, then cover with cling film or laminate to stop it washing away in a shower. Keep it simple – "Eggs For Sale" says it all.

In smaller letters you should put a message like "Just Knock on the Door" or "House up the Path, Just Knock on the Door". Many people are reluctant just to walk up a private drive and knock. We know it seems a bit unnecessary but trust us, it tells people they are welcome and have come to the right place.

Another good tip is to have a second sign under the top one which is on hooks and can be changed saying "Eggs in Stock" or "Sorry Sold Out Today". This actually does two things: it tells them they are not wasting their time coming up your drive; or if they see "sold out" it tells them your eggs are really fresh and popular.

Be careful about claims over your eggs. Avoid using descriptive terms like "*organic eggs*" or "*free-range eggs*". These are legal terms and you may need to prove your status legally. If someone asks, offer to show them your hens. Once they see your hens and you explain why you hate battery keeping you will have a customer for life and probably a new friend.

If the sign doesn't raise enough interest to move your surplus eggs, perhaps you're out of the way and off the beaten track; you'll need to build up a customer base. *The best method is word of mouth*. If you have a load of extra eggs and you're not going to be able to sell them, giving some away can be a good investment in future sales. If someone buys a dozen eggs, give them an extra half dozen and tell them to tell their friends.

Advertising in newspapers is certainly not cost effective for someone selling a few eggs from home but a few leaflets in the locality will help. If you're hard to find, then a little map can work wonders, along with a phone number if they get lost.

## Stock Control and Storage

Whether you're selling eggs or not, you need to keep some control over your eggs or you won't know which are fresh and which are approaching the end of their storage life.

Eggs are best kept in a fridge at around 5°C and taken out a few hours before use to allow them to come to room temperature. You can keep an egg in the fridge for a surprisingly long time. We've found them perfectly edible after a couple of months.

These are eggs for our own consumption. We wouldn't dream of selling or giving away eggs over a couple of weeks old, just in case they had gone over.

Eggs will store at room temperature but, simply put, the higher that is, the shorter their storage life. In any case, you need to use or sell the oldest ones first.

It really doesn't matter on a domestic scale to a day so we just write the week on an egg box and keep them in that. As the box empties, you cross out the week and start the current week's box. It really can't be simpler.

If, as sometimes happens to the best of us, you find some eggs in an obscure place in the garden then don't assume they're fresh or even all laid on the same day. Keep them separately and crack them into a cup before use, just in case they're older and have gone off.

There is a trick you can use to estimate the age of your eggs. As the egg ages, the size of the air space grows. Put your egg into a basin of water. If it just lies there, then it's not got a large air space and is pretty fresh. If it starts to lift so the rounded end is higher than the pointed end, it is less fresh; and if it floats, it is beyond eating.

If you get an egg that floats, you won't be able to resist checking that it really has gone off. Take it outside to crack it, because the stink of a rotten egg seems to linger inside for ages.

Because our birds lay more in the summer than the winter, we often have that glut of eggs that need long-term storage. In the past, various methods were used to store eggs for months but they are all a lot of fuss and not very reliable. For the sake of completeness, we'll cover them for you.

**Waterglass** was commonly used. Waterglass is now pretty hard to get but you may find a chemist who will be able to order it in for you. Its chemical name is sodium silicate. Mix the chemical with water, one part waterglass to nine parts water and put into a food-grade plastic bucket with a lid or, to be totally traditional, a glazed earthenware pot with a lid. Eggs immersed in the solution will keep for up to six months. However, if the egg has a hairline crack or other damage, it will go off. Always crack separately any eggs stored in this way before using, just in case.

Storing eggs in a strong **brine** solution extends the storage life but the eggs absorb the salt which not only makes them taste salty but, as we now know, too much salt is bad for you.

**Drying eggs**, as in the infamous wartime dried eggs imported from the USA, is not something you can do on a domestic scale.

**Painting eggs with gum Arabic** is another traditional storage method. Each egg has to be individually painted, left to dry and then upended and the other half painted. It takes ages, is messy and not very effective. Happily we now have freezers.

**Freezing eggs** is very easy. You can't just put your whole egg into the freezer, because liquid expands as it freezes and you'll end up with a mess. The answer is to use deep ice-cube trays. Take your egg and separate the yolk from the white. If you're doing a lot, an egg separator which is available very cheaply in the shops will save time. Put the yolk into one cube and the white into separate cubes. Usually you'll find that an egg will make one yolk cube and two white cubes, which is very useful when you come to use them.

Once frozen, place the tray in an airtight bag to stop desiccation in the freezer. For long-term storage, over six months, they will maintain their flavour better if you whisk a little salt into the eggs. Obviously they're only able to be used for savoury dishes if salted.

For use in sweet dishes, like cakes, replace the salt with a little castor sugar. Do label the bags or you could be in for some interesting dishes. Adding salt or sugar also helps to prevent a skin developing on the eggs during the freezing process.

Another method is to just crack two or three eggs into a bowl, whisk them together and then freeze them in a plastic freezer bag. We have a couple of square freezer boxes, one contains two egg packs and the other three egg packs. Check the cake recipe and pull out the appropriate pack.

The eggs, once defrosted, can be used for any cooking purpose that you could use fresh eggs for. However, you'll discover that soufflés may not rise as well or omelettes are not quite as fluffy as you want.

### Pickled Eggs

The last method of storing eggs is to pickle them. Pickled eggs are no longer as popular as they once were, which is a shame, but as with a lot of foods, once people have tried the real homemade product their attitude changes.

It's best to use eggs that are around two weeks old as they'll be easier to shell.

To hard boil the eggs, place into a saucepan of cool water and then slowly bring to the boil, stirring the eggs gently during the first few minutes of boiling. This helps ensure the yolks are centralized.

Boil for about 8–10 minutes and then plunge into cold water immediately on removing from the pan to prevent the yolks getting a black ring around them.

Once cool, shell the eggs and pack into a clean, sterilized, wide-necked jar. A 0.75 litre, or even 1.5 litre, Kilner-style jar is ideal as it can be re-sealed as you use up the eggs. Then pour hot spiced vinegar over the eggs, close the lid, label and keep in a dark store cupboard.

Sterilizing the jars is easy enough: scald with boiling water and, having removed any rubber ring seals, such as those you get with Kilner-style jars, place in a cool oven for 15 minutes or so and the job's done.

The flavour will, of course, vary depending on the spiced vinegar used. Some like to use a white vinegar, others a dark malt; and even cider vinegar can be used. You can buy ready-made spiced vinegars but it's fun to experiment and discover your perfect pickled egg.

The important thing with the base vinegar is to ensure it has a minimum acidity level of 5 per cent. This should be stated somewhere on the bottle; if not, don't buy it. The choice of spices is up to you. Perhaps easiest is to start with a specific pickling spice mixture and then adjust as you become more adept.

A good standard mild pickling vinegar would use the following spices:

¼ oz (7 g) cinnamon
¼ oz (7 g) cloves
¼ oz (7 g) mace
¼ oz (7 g) whole allspice berries
6 white peppercorns

If you want to increase the kick, try adding dried root ginger, mustard seeds or even chilli peppers. Be careful though, you don't want to overwhelm the delicate flavour of the eggs.

Place the vinegar and the spices (tied in a muslin bag) in a heatproof basin and stand the basin over a saucepan of water. Cover the basin with a plate or the flavour will be lost with evaporation. Bring the water in the pan to the boil and then remove it from the heat. Set aside for 2–3 hours to allow the spices to steep in the warm vinegar. Strain the vinegar, re-heat but do not boil and pour on your eggs.

People usually associate pickling with hens' eggs but ducks' eggs pickle just as well and pickled quail eggs make a lovely unusual present or even a very saleable product. With pickled quails' eggs, only ever use a mildly spiced vinegar though.

# 16

# KEEPING POULTRY
# FOR MEAT

At one time it was quite common not only to keep a few hens in the back garden for their eggs but also to raise some meat birds. Remember that half the eggs hatched will turn out to be cockerels and unwanted. So they would be raised to end up in the pot.

The hens would also go into the pot once their productive egg-laying years were over. They would be relatively scrawny and tough but that's what slow cooking, soups and stews were invented for. That French classic dish, coq-au-vin is basically a peasant dish for using up tough old birds. Remember that wine is, or at least was, dirt cheap in France. The acidity of the wine would soften the meat and the long cooking in a pot on the kitchen range would produce a wonderful dish.

Commercially chickens have been bred to be either egg-layers or meat birds. With the egg-laying breeds (hybrids), the chicks are sexed and separated within a day or so of birth. The "useless" males are then gassed and become a waste product.

The fate of the meat birds is hardly much better. Over 90 per cent (some put the figure as high as 98 per cent) of the 800 million meat chickens consumed annually in the UK are raised in broiler houses. These are just huge sheds where thousands of birds are crowded in with hardly enough room to move and just sit, eat and put weight on.

Their treatment is hardly what you'd call humane. One good example we saw at an agricultural show was a machine designed to collect the birds from the shed. Imagine a huge go-kart with a sort of rubber spiked roller brush at the front.

Apparently it's an efficient way of picking up a large number of hens from the broiler house.

Selective breeding has resulted in chickens with a phenomenal growth rate. Most broiler chickens are slaughtered at just six weeks old. So fast do these poor creatures grow that their bone structure can hardly support them. Many suffer from leg disorders due to this; the bones can break just from the strain. Think of putting the torso of a 20 stone man onto the legs of a three-year-old.

You can even see a result of this in the packaged chicken on the supermarket shelf: hock burn – a discoloration of the hock caused by them sitting in their own excrement. This can happen, although less often, with free-range birds. The reason here is that, whilst the birds are raised according to the legal requirements to be described as free-range, the sheer numbers involved mean that many birds don't actually go out of the sheds and those that do tend not to go too far, hence the area around the doors is pretty crowded.

One way you can tell if the bird has been truly free-ranging, as against just brought up in conditions that meet the letter of the law, is the conformation. Most of the meat on a broiler house bird is on the breast but true free-rangers have much more on the legs and less on the breast in proportion.

If those details have not put you off eating cheap mass-raised chicken, here's another reason. Chicken has been promoted as a healthy meat, low in fat and high in protein. That's certainly true for traditionally raised breeds but the birds raised in the broiler house just sit and eat, putting on weight. And just as if we did the same, the result is fat. Most of our bought chicken is now a high fat meat and we'd argue it is no better for you than any other meat. Some would say worse.

Incidentally, all the above equally applies to other poultry. Ducks and turkeys are intensively raised in very similar conditions. Although the vast majority of producers adhere to the regulations, there is an effect on the people who work in the industry. They cease to see them as living creatures and develop a certain callousness. The extreme result was shown by the awful behaviour of employees on a turkey farm playing baseball with live birds.

As we touched on above, you can buy free-range birds which are better but because of the volumes not, in our opinion, all that you may expect and believe. The best way to buy quality meat birds is direct from a small-scale producer or a butcher who can demonstrate provenance and knows his supplier and can confirm he meets your standards.

Raising your own birds at home for meat is quite easy, economic and you know for certain that they have been decently raised, had a good albeit short life and are a good lean, low-fat, high-protein food for your family. They taste great as well.

Those who grow their own vegetables and fruit will tell you how much better the flavour is than shop bought. It's just the same with chicken and poultry, more so if anything.

Before you start raising meat birds, you need to be sure you can actually cope with the psychology of it. Yes they've been raised to a high standard, well looked after and undoubtedly had a better life than any bird raised in a broiler house but actually killing a living creature is not something that comes easily to us.

You can raise ducks for the table but our plan to raise a couple of Aylesbury cross drakes to eat came to nothing. Not only did they decide we were their parents, we decided they were our babies! Naming your meat birds is not a good idea either, as we said earlier. You need to distance yourself from them. Oh well, the drakes Bumble and Bee are great pets and living to a happy old age.

You can dodge around the attachment problem if you get together with another poultry keeper. You give your meat birds to her and she gives her meat birds to you to do the awful deed.

There are, of course, other backgarden birds that can be raised for meat. Ducks, quail, geese and turkeys can all be raised for meat alongside your egg-laying flock. With ducks, in particular, cross breeding two different large fowl seems to create a larger meat bird if you select the male offspring for the table.

This also means that, when hatching, by selecting the male hatchlings for the table you can avoid having to keep too many cock birds in your garden. Because of our failure with table ducks, we'll concentrate on raising chickens for the table!

We are talking here about raising birds for your own consumption, not for sale. If you sell the meat, there are various legal requirements you need to follow.

**What Breed?**
One of the first things to consider is what breed of chicken you wish to rear for the table. There are a number of options here.

Firstly, if you are hatching chicks to supplement your current egg-laying flock you can just grow on your cockerels for meat. Cock birds are notoriously hard to re-home unless they are high-quality breeding birds. This method is best suited to larger breeds that will have more meat on them. A bantam breed is not going to make a good table bird size-wise, although the taste would be just as good as home-reared larger fowl.

Secondly, you could pick to hatch a specific dual-purpose breed of chicken to replace egg-layers and provide large meat cock birds for the tables. Breeds such as the Sussex, Dorking or even the Transylvanian Naked Neck (which also saves you some plucking after dispatch) are good dual-purpose breeds. The hens are good large egg-layers and, although not as prolific as hybrids, they have a longer laying life and are consistent layers even through the winter months.

The cock birds will grow at the same rate as the hens and may be kept together, but for a fuller weight bird separation and finishers feed are recommended once the hens are moved onto layers pellets.

Most cock birds will be ready for the table by 18 weeks, or perhaps when they first crow, which can be from 16 weeks onwards for a smaller carcass or can be left to 28 weeks for a nice large and full-flavoured bird.

Thirdly, you could raise table-specific breeds (broilers) such as Hubbards, the Cornish-Rock or the Ixworth. The main problems with breeds specifically designed for meat is that they grow incredibly fast and sometimes struggle to take advantage of free-range facilities even when they are available as they find it difficult to move great distances.

Economically this would be the cheapest method of raising

meat for the table, as the birds will be ready for eating sooner than a dual-purpose reared cock bird. A specific meat bird can be ready for the table with a dressed weight of 2–4 lb (1–1.8 kg) within 6 weeks.

A broiler chicken raised at home, if provided with space and stimulus, will have had a far happier life than one raised in a commercial chicken farm. The free-range chicken you buy from the supermarket will be from one of these fast-growing breeds, so you will, by giving them space to choose to roam, be giving them at least the same standard of life as a free-range commercial bird.

Our feeling is that these breeds are not best suited for the home producer though. The problems of their fast growth rate still apply even in the best conditions. With these broiler breeds, the taste in our opinion is not as good as a well raised free-range cock bird.

If you are raising birds to make sure your meat has had a happy life before being culled, the dual-purpose breeds are a better option. A cock bird raised free-range will still cost less than the equivalent supermarket or butcher's free-range bird, although a few pounds more than the broiler chicken.

## Looking after Table Birds

You should give your meat birds the same standard of living conditions as you would your egg-layers. They need at least 1 ft² (0.1 m²) of coop space and 1 m² (11 ft²) of run space per bird as well as entertainment such as dust baths. Feed and water should be supplied ad lib and should be topped up and freshened on a daily basis.

For dual-purpose breeds or general cock birds, you can feed them along with their sister hens until 14–16 weeks of age on un-medicated growers pellets. Usually hens are moved to layers pellets by 16–18 weeks of age and at this stage you can choose one of two ways to feed the birds intended for the table.

The first feed method is to allow the cock birds to eat the same layers feed as the hens. This will lead to a slower growing bird that is still full of flavour and will mean you can continue to run the birds together and not have multiple types of feed.

If doing this we would recommend for the 3 days before you are going to cull, moving the cock birds onto a diet of corn to improve their flavour. The birds will be table-size by 22 weeks of age for a smaller carcass.

The second method is to separate the birds and move them onto finisher pellets. This feed helps to increase the body mass of the bird quicker and will mean you have a larger carcass or will be able to cull the bird sooner. We would again recommend for the 3 days before you are going to cull moving the cock birds onto a diet of corn to improve flavour.

For meat-specific breeds, the timetable differs a little. You can use usual feeds in the form of chick crumb, growers pellets and then finishers pellets or you can use feeds that are aimed at obtaining a higher body weight sooner such as broiler starter and broiler growers.

The bird should be moved from starter/chick crumb at 3–4 weeks of age to a growers pellet and then at 7 weeks of age to a finisher pellet. You will then have a bird ready for the table from between 8–10 weeks. You will notice, however, that although the birds eat for a shorter period than a dual-purpose breed, they eat a far increased amount. You should therefore give larger feeders and drinkers and always top up the food and water daily, if not twice daily, if possible.

All birds need to be kept under heat after being hatched until at least 4 weeks of age. Obviously if rearing with a broody hen, she does this for you, but with broilers and incubator-hatching you need to make sure you keep them warm and clean to decrease any losses at the early stages.

## The Economics of Raising Your Own Meat Chickens
The cost of raising your own birds depends on a number of factors. You will need the basic equipment of an incubator (or broody hen), heat lamps, a coop and run, and feeders and drinkers. This assessment assumes you already have the equipment and is based on the cost of chicks and feed alone for broiler and dual-purpose breeds on a free-range basis.

The first cost comes from whether you choose to hatch your

own eggs or whether you purchase day-olds. If you hatch your own, six hatching eggs will cost on average £6 for a broiler breed or between £10–18 (so let's say £14) for dual-purpose breeds (based on Light Sussex).

If you hatch, you may only get 4 of those eggs hatching successfully which gives you an average cost of £1.50 for the broiler chick and £3.50 for the dual-purpose chick if using a broody hen to carry out the hatch for you. (You also need to remember that 50 per cent of your dual-purpose are likely to be hens for your laying flock which is why they cost more to buy.)

If using an incubator and the heat lamp, you will need to add another £0.25 per chick in running costs. If buying day-old chicks you will be looking at 4 broiler chicks, costing an average of £1.65 each, and 4 dual-purpose breeds, costing an average of £3.50 per chick.

| Type | Day-old Hatch | Day-old Buy | Using Heat Cost |
|------|---------------|-------------|-----------------|
| Broiler | £1.50 | £1.65 | £0.25 |
| Dual-purpose | £3.50 | £3.50 | £0.25 |

As you can see, the initial cost for a dual-purpose chick (based on Light Sussex) is more than double than a broiler. However, if you are also raising hens, this is a good investment, as an 18-week-old POL Light Sussex hen would cost you an average £16 to purchase and probably 50 per cent of the hatch will be female.

Inevitably there are some losses when raising chicks, but we'll ignore them on this small scale.

The next cost is the cost of feed for the growing chicks. Let's say that an average 20 kg (44 lb) bag of non-organic feed costs £6.50 or 32.5p per kilogram. The price varies but this is a fair estimate for a home breeder at the time of writing.

A broiler chick hybrid has been bred to grow and will eat far more than a pure breed. So while they eat more, it's for a shorter time and since they can't move around it's efficiently converted into body weight even if it is fatty. This is what makes them so attractive to the commercial producers.

| Feed Consumption of Typical Broiler Bird | | |
|---|---|---|
| | Feed Consumption (kg) | |
| Age in Weeks | For the Week | Cumulative |
| 1 | 0.14 | 0.14 |
| 2 | 0.33 | 0.47 |
| 3 | 0.52 | 0.99 |
| 4 | 0.73 | 1.72 |
| 5 | 0.95 | 2.67 |
| 6 | 1.17 | 3.84 |
| 7 | 1.42 | 5.26 |
| 8 | 1.55 | 6.81 |

That means that each broiler chick will cost about £2.20 to feed up to weight for culling.

A dual-purpose bird will eat less, exercise more and take longer to reach weight. The result is a far better table bird with less fat and more flavour but the costs are higher. For a commercial producer, the deciding factor is the time the bird takes. They can get twice the use of their housing and manpower, thereby halving their costs as well as reducing the food bill.

| **Feed Consumption of Typical Dual-Purpose Bird** | | |
|---|---|---|
| | *Feed Consumption (kg)* | |
| *Age in Weeks* | *For the Week* | *Cumulative* |
| 1 | 0.09 | 0.09 |
| 2 | 0.27 | 0.36 |
| 3 | 0.41 | 0.77 |
| 4 | 0.54 | 1.31 |
| 5 | 0.54 | 1.85 |
| 6 | 0.54 | 2.39 |
| 7 | 0.68 | 3.07 |
| 8 | 0.68 | 3.75 |
| 9 | 0.68 | 4.43 |
| 10 | 0.68 | 5.11 |
| 11 | 0.68 | 5.79 |
| 12 | 0.68 | 6.47 |
| 13 | 0.68 | 7.15 |
| 14 | 0.68 | 7.83 |
| 15 | 0.68 | 8.51 |
| 16 | 0.68 | 9.19 |
| 17 | 0.68 | 9.87 |
| 18 | 0.68 | 10.55 |

That means that each dual-purpose chick will cost about £3.40 to feed up to weight for culling.

Therefore, the average cost per broiler chick being home-reared from a hatching egg, including using heat and taking into account the initial chick cost and feed is £3.95.

The average cost on the same basis of the dual-purpose chick is £7.15, but it's a far higher quality product.

Compare this to the average cost of a shop or butcher's bought free-range chicken weighing 2 kg at £8 and you see that the broiler saves you nearly 50 per cent and the dual-purpose still gives you a saving of £1 per bird!

Of course, if you breed your own birds that saves the cost of buying in either eggs or day-old birds. But that does require a cockerel and, as we've said elsewhere, that can be a major problem in a suburban setting.

Regardless of costs, you have the satisfaction of knowing where your meat came from, being certain it was humanely reared and that makes for a very special chicken dinner.

## Killing a Bird

There are three situations in which a bird will be culled: unwanted, meat or illness and injury. It's a sad fact that half the eggs laid will contain a cockerel and since they don't lay eggs they are mostly unwanted or destined for the table. With hybrid-laying hens being bred to lay with a minimal amount of weight gain, most of the commercial hatcheries will kill the unwanted male chicks at one day old. All 40 million of them. A sobering thought.

The vast majority are gassed using either a high concentration of carbon dioxide, which is cheap and efficient but can cause distress. A more humane gas but more expensive is argon or nitrogen. The birds die of lack of oxygen but don't realize there is no oxygen available. John can attest that breathing an inert gas will not cause distress having nearly made himself unconscious being silly with helium in his youth.

With birds raised for the table, the method of killing has to comply with *The Welfare of Animals (Slaughter or Killing) Regulations 1995* (WASK '95) (as amended). This states that slaughterers have an obligation to ensure they do not cause any "*avoidable pain, excitement or suffering to any animal*".

Generally, this is taken to mean in practice that poultry are rendered unconscious by passing an electric current through their brain by means of electrodes attached to the head prior to slaughter or in large commercial operations by an electrical waterbath.

Slaughter is usually carried out by cutting the large blood vessels in the neck of the bird or by inserting a knife into the mouth and bleeding the stunned bird. Death occurs within a minute but causes no pain as the bird is already unconscious.

A small-scale electric stunner will cost from around £350–£600 new but may be obtained cheaper second hand.

Another method available for relatively small-scale producers is the Cash Poultry Killer. This is a bolt gun type of device which causes the death of the bird. After being killed, the bird still needs to be bled as for electrical stunning to comply with the above legislation. The "gun" uses .22 cartridges. A new one, with 100 cartridges, costs around £550.

The traditional method of dispatching small fowl has been neck pulling. If carried out correctly, it will cause extensive damage to the brain and render the bird immediately unconscious with complete death occurring within seconds. The problem with this method is that it requires training to be performed correctly.

Having discussed this at some length, we don't feel it appropriate to try to explain exactly how to pull a bird's neck. You need to find someone to show you how to do it in the flesh and to be on hand if things go wrong. Even having been trained, it's a good idea to keep a sharp cleaver or an axe to hand in case things go wrong.

You can purchase manually operated humane dispatchers from around £30. The hand-held version is similar to a pair of pliers and is designed to break the neck. A wall-mounted version is also available for a little more money. We believe that this is more effective as it severs the spinal cord. The Humane Slaughter Association contends that humane dispatchers crush the neck and are, in fact, not humane at all. Personally we feel that death occurs so quickly with humane dispatchers that they are humane.

The final method is decapitation. Once again the Humane Slaughter Association states that this is not humane as a study has shown some brain activity continues for 30 seconds afterwards.

The immediate effect of decapitation is to drain blood from the brain, a massive loss of blood pressure. Having once collapsed due to low blood pressure, John cannot agree that it is inhumane. The effect was a painless swift loss of consciousness and that was a minor loss of pressure in comparison to decapitation.

If you find yourself with a badly injured bird, perhaps following a fox attack, then you can attempt to find a vet or relieve the suffering yourself. A vet is unlikely to be able to get to you or you get to the vet within half an hour and that is a long time to allow a fatally injured bird to suffer.

We would suggest the use of an axe or cleaver to decapitate the bird or even a hammer blow to the head. Remember that the purpose here is the immediate relief from pain and we're looking at the lesser of two evils.

Do be calm and decisive. An extra second or two will make no difference. Take your time to do the job properly. Be careful. You don't want to get your hand in the way of the axe.

Following death, the bird's wings will flap frantically and the legs will move for a few moments. This is purely a muscular reaction, not a sign of life. It is distressing and concerning to see, but do understand that your bird is now beyond suffering.

# GLOSSARY

**ACV**                     Initials of Apple Cider Vinegar,
                            see page 69.

**Air Sac**                 A pocket of air found in the egg,
                            see page 149.

**Albumen**                 The egg white.

**Autosexing Breed**        A breed that allows sexing of
                            chicks at day-old, based on
                            feather coloration.

**Avian Flu**               A disease/illness. See page 98.

**Bantam**                  A miniature fowl about a third of
                            the size of large fowl.

**Broody**                  A hen or other female bird that is
                            preparing for and/or sitting on
                            eggs. See page 104.

**Bumblefoot**              Infection of the foot. See page
                            89.

**Candling**                Process of checking the stage of
                            development of a hatching egg.
                            See page 108.

**Caruncles**      Lumpy flesh found on the faces of Muscovy ducks and drakes.

**Chicks**      Young chickens.

**Cloaca**      Situated behind the vent where the internal canals are located in poultry.

**Coccidiostat**      Medication used to treat coccidiosis. See page 88.

**Cockerel**      Male chicken.

**Comb**      Fleshy crest on top of a chicken's head.

**Crop**      Where food is collected before passing through the gizzard for digestion.

**Drake**      Male duck.

**Dual-purpose**      Birds that are good for both egg-laying and for meat.

**Ducklings**      Young ducks.

**Dummy Eggs**      Also known as pot eggs. These are pottery or wooden eggs used to encourage egg-laying in certain places, to encourage broodiness or to break egg-eating.

**Egg Tooth**      Tip of a chick's beak used to pierce the eggshell during hatching.

**Flight Feathers**      The primary and secondary feathers of the wing in poultry.

**Gander**            Male goose.

**Gizzard**           Part of the digestive system where grit is stored for breaking down of food.

**Goslings**          Young geese.

**Hybrid**            Cross of two pure breeds for a specific purpose, such as egg-laying or meat birds.

**Layers Pellets**    Compound feed fed to hens from point of lay (18 weeks), also in mash form and follows on from chick crumb and growers mash or pellets.

**Moult**             Shedding of old feathers and the subsequent regrowth of new feathers.

**Oil Gland**         Located at the base of the tail in water fowl, it produces water-repellent oil used while preening.

**Pin Feathers**      Newly emerged feathers after moult.

**Point of Lay**      Female poultry aged 18 weeks and over who are about to come into lay for the first time.

**Poultry Saddle**    A saddle used for hens during the mating season that helps prevent claw and spur damage by the cockerel.

**Poultry Spice**     Food supplement – see page 67.

**Poultry Tonic**           Liquid version of poultry spice.

**Primary Flight Feathers** The first ten flight feathers on the
                            bird's wing.

**Pullet**                  A young hen under 12 months of
                            age.

**Roost**                   The perch or bar on which birds
                            sleep. To roost means to sit on the
                            bar.

**Sex Curl**                The curl of the drake's tail
                            feathers in mallard-derived
                            drakes.

**Spur**                    The horned growth located about
                            the foot predominantly on male
                            chickens.

**Utility Bird**            A breed that is useful for both its
                            egg production and meat-bird
                            abilities.

**Vent**                    Orifice where both eggs and
                            excretions are passed.

**Wattles**                 Hanging fleshy lumps on either
                            side of the face found in some
                            poultry.

# FURTHER INFORMATION

As your interest in poultry keeping grows, you will naturally want to find out more. There are specialist books available on everything from diseases to breeds, which are useful. However, there are times when you may have a problem or just need a second opinion. Getting help from other keepers is invaluable.

You'll discover that the poultry community is full of wonderful people who share their knowledge and advice freely. You can meet them at poultry and agricultural shows but in today's busy world you may not have the time to attend them. This is where the internet comes into its own.

On our website **www.allotment.org.uk**, we have a large poultry help section, lists of breeders, poultry keeping courses, etc. We also have a large chat forum where you can virtually meet keepers, share your experiences and get advice on any problems you may have.

# INDEX